# FIX OUR INFRASTRUCTURE

How the city of Atlanta, Mayor Shirley Franklin, and her administration succeeded.

by
Gilbert Clay

ISBN: 1482659972
ISBN 13: 9781482659979
Library of Congress Control Number: 2013906716
CreateSpace Independent Publishing Platform
North Charleston, South Carolina

# Preface

In 2000, I had the opportunity to be interviewed by Khafra Engineering, which was looking for inspectors to start work on a  wastewater and stormwater separation project with the City of Atlanta. When I returned home that day from the interview, I received a call from H.J. Russell Construction Co., which was also starting infrastructure work on another project-(The Villages at Carver, Phase I, a renovation), to come in immediately for an interview. Wow, I thought, two interviews in one day—how rare. I chose one over the other for more money. I was laid off from that project one year later and thought about the decision I had made to switch job opportunities. Prior to that, I was an Inspector working on MARTA projects, the last being the Sandy Springs Station Phase I, relocating utilities, installing new water and sewer piping and manholes, installing the core and shell for the underground station, and renovating the Perry Boulevard bus facility for the Olympics in 1996 to receive the new buses running on compressed natural gas. My first construction inspection work in Atlanta was during the Grady Memorial Hospital renovation and expansion, where I served the chief engineer for the hospital as a construction liasion between contractors involved in the construction work and hospital's operations team during the last two phases of new construction, which included the new clinic building.

After Grady, the minority company I worked for obtained a contract with CH2M Hill to help with upgrading the City of Atlanta's outdated storm maps. We had a small team that went out and tallied and assessed the conditions of all the storm drains and reported that information on data sheets back to CH2M Hill. The company in turn gave that information to the city's GIS department to create city identification numbers and coordinates for the structures.

Several years passed and I got the opportunity to work on the last year of the Greensferry Sewer Separation project as an inspector. After Greensferry closed out, word of mouth landed me an interview in the North Area field office of the SSES project for the City of Atlanta.

So here I am, over ten years have passed and I've returned working on the SSES projects for the City of Atlanta and writing about what I walked away from. If I only knew then what I know now.

Infrastructure work is on the top of the president's list because these projects are "shovel ready" and, when put on the books as capital-improvement projects and passed by city councils, they put people back to work, boost the economy, and their impact greatly improves the quality of life for residents, communities, the environment, and the states.

Infrastructure is not a foreign subject even though from the surface you cannot see what is going on. It's the best work you will never see. This book is not about a bunch of formulas, equations, symbols, or calculations that will bore you after several pages but an easy read about what goes into fixing a major city's infrastructure. I am a transparent person when it comes to information; give it to me in a way I can understand it.

This research started out as an assignment for a graduate class. I chose to write about the "rehabilitation of aging sewer piping" since this is a subject that I knew I could write about. After completing the class, I decided to challenge myself to do more research, collect more information after graduation, and turn this into a book. What about a title? Looking around me at work, I saw a lot of people planning, and I witnessed major rehabilitation construction on the city's infrastructure all over town. This was my epiphany.

## This book's purpose is threefold:

*One-* to address how the City of Atlanta assessed the problems, found solutions, fixed its infrastructure, and laid out a path for the future.

*Two-* to highlight the achievements of Mayor Shirley Franklin and the City of Atlanta's efforts for the years starting from January 2002 to December 2009 on completing the mandated work imposed on Atlanta's consent decree(s) to fix its water and sewer infrastructure work and formulate a strategy to assess the city's CSO (combined sewer overflow) problems that resulted in the creation of the department of watershed management that started a chain reaction of other agencies for the city to look at masterplanning and to go green.

*Three-* to give officials with no working knowledge of infrastructure the ability to understand what it took to achieve such a vision, discuss with their constituents how best to proceed to fix their cities' aging and crumbling infrastructure, and formulate a strategy that comprehensively covers the costs over time.

Lastly, this book will give educators the ability to discuss with students how important it is to improve the aging infrastructures of cities and come up with better design techniques and construction methods to deal with the very complex and unique problems that citites have.

A final chapter, "What's next?" addresses the transportation projects Atlanta and the Georgia Department of Transportation are tackling to keep up with demand on its roads.

I have, to the best of my ability, attempted to describe from my own experience, my conversations with city workers and managers, my readings of magazine articles, and my Internet searches what Mayor Franklin and her efforts created to define the structure, the people, the cost, the maintenance, and the operations to transform a city from a mayor's vision.

The information presented here should be a blueprint for other cities to emulate, to learn from, and to start the process of getting infrastructure projects (capital improvement) on the city council's agenda if they truly want to "fix their infrastructure" and make it a world-class system of the kind created in the City of Atlanta under the pointed direction of Mayor Franklin . The process may not be as involved as was the case with the City of Atlanta, but the results should be the same. The complexity and scope of the process in each city is different based on the city's size. Either way, a Sanitary Sewer Evaluation Survey should be the first step.

My hope is that this book gets in the hands of every elected official who lives in a city, county, or state in desperate need of repair where the officials can implement a department of watershed management such as the one created in the City of Atlanta. If your city is not as big as Atlanta, the department of public works should be able to set aside budgeting for capital improvement projects like booster pump stations, SSES maintenance, valve replacement, meter replacement, GIS mapping, storage tunnels, upgrading water and sewer treatment facilities, or providing areas for green space.

*Note:* By <u>NO</u> means is the information presented here complete. The information is described from my perspective and does not reflect all the events that went on during Mayor Franklin's time in office. The information presented gives only a glimpse. I'm sure if I had had a sit down with Mayor Franklin, I could have obtained more detailed information that would have changed how I approached this research.

Enjoy.

## Acknowledgments

To all those who have given me their personal insights on this topic.

## Dedication

To Mayor Shirley Franklin, her administration, and the City of Atlanta for taking on the task of moving Atlanta's infrastructure forward and succeeding. She truly took this task to heart to make Atlanta _the_ premier city, a model for infrastructure work and "Best in Class."

# Table of Contents

# Introduction

## Why Study Infrastructure?

There are different platforms of infrastructure, such as water, sewer, oil and gas, telecom (aerial and underground), roads and bridges, transportation, airports /air transit, banking/finance, information technology, safety /security and more recently, solar energy.

Wikipedia defines infrastructure as referring to the "technical structures that support a society." For the purpose of this reading, we will refer to infrastructure as water, sewers, and transportation.

One-hundred-year-old pipes that are aging and deteriorating pose many risks, including lead contamination, root intrusion, crushed and collapsed piping, reduction, and restricted volume of flow through pipes.

The pipes that are in the ground today were laid at different times, were made of different materials and manufacturing techniques, and have different life expectancies.[1] Cast-iron pipes were laid in the late nineteenth century and have an average life expectancy of 120 years. Ductile iron pipes were introduced in the 1950s, a time of rapid population growth, and were marketed as an improvement over their cast-iron predecessors. The life expectancy of ductile iron is fifty to seventy-five years. In addition to traditional cast-iron and ductile pipes, the last half century also

---

1      http://www.scribd.com/doc/88877493/Bonner-Cohen-Fixing-America's-Water-Infrastructure

saw the expanded use of pipes made from corrosion-resistant polyvinyl chloride (PVC).

Manholes and sewer structures that are aging and crumbling are outliving their useful life and need to be rehabilitated or replaced.

Restricted flow results in the buildup of grease, oil, and fat. It also causes sewage to seep into the ground where pipe joints have disconnected themselves. Repairing, replacing, and upgrading large- and small-diameter concrete piping for infrastructure in metropolitan cities is an interesting subject now in the twenty-first century because of their present condition. Removing harmful lead contaminants from deteriorating pipes, increasing (upsizing) pipe diameter to allow for better flow and future growth, correcting aging pipes and manhole structures, and assessing the current condition of an aging system are the issues that led to this study. What influences the deterioration of sewers and piping is the risk or likelihood of collapse.

There are three stages of collapse:

Stage 1: an initial defect enables the deterioration process to continue.

Stage 2: the deterioration process continues in/and/or behind the sewer wall.

Stage 3: collapse due to weakened walls.

## 1. Significance of the Study

This book will look into the City of Atlanta's mandate for fixing its infrastructure, what the city , Mayor Franklin, and her administration did to "fix our infrastructure" with emphasis on sewer assessment, maintenance, state-of-the-art repair methods, building new tunnels that transport waste to treatment facilities, allocating city green space throughout the city and its parks; how the city built a state-of-the-art testing laboratory that tests and treats water; how it forged partnerships with business and community leaders; and how it created a resurgence of the city, positioning it as a model for infrastructure reform.

Here is a little bit of the history of the Clean Water Act and some background on Atlanta.

History has always recorded and documented important milestones so future generations can know of our past.

The important information that was collected for this book (from various sources) shines the light on the mandate set upon the City of Atlanta, and on then Mayor Shirley Franklin as Atlanta's new mayor to perform a task that transformed a city.

Research from articles, letters, and transcripts, and archived information from websites pertaining to Atlanta's sewer and water issues were considered valuable information to validate dates and sequences of events.

An article[2] by Hamida Kinge states that most buried sewer and water infrastructure in the United States are more than one hundred years old. Starved of the funding required to sustainably manage, maintain, repair, and rehabilitate their assets, the vast majority of water utilities across the country are watching them crumble. Upgrades and maintenance are astronomically expensive and extremely disruptive, and since utilities have very little knowledge of water-main conditions, it can be hard to find leaks before they become breaks. Each year there are 250,000 water-main breaks— close to 700 a day. The article further states that even today, the vast majority of cities and communities lack the technology needed to monitor and maintain buried assets.

This study's importance arises from the need to survey, assess, and maintain sanitary sewer systems and water-distribution systems that are aging and

---

2    "What's on Tap: America's failing water infrastructure," AmericanCity.org

deteriorating to a point where serious environmental risks cause harm to the population. The EPA asserts flow surging from manholes into streets that travel into storm drains, and makes its way to the river pose serious health hazards to humans and fish. A need to separate the combined sewer system is addressed, as well as a need to repair and rehabilitate aging water lines that are corroded and which cause biohazards for drinking, and water loss from the system through infiltration and exfiltration. Aging water infrastructure can pose serious risks to public and environmental health. The need to repair broken/leaking water valves and meters are addressed. Spills and dumping into the streams, creeks, and rivers have led to protection by stream buffers, and training and certification of personnel. Treating wastewater and storm water and returning it to the environment are always vital to a sustainable world.

Aging infrastructure and outdated sewer systems have brought about the chronic discharge of raw sewage into US surface waters.[3] The EPA estimated in August 2004 that the volume of combined sewer overflows (CSOs) discharged nationwide is 850 billion gallons per year. Sanitary sewer overflows (SSOs) result in the release of as much as ten billion gallons of raw sewage yearly. The EPA estimates that there are at least twenty-three thousand to seventy-five thousand SSOs per year, and the agency has been getting serious about CSOs and SSOs. According to the *New York Times*, between 2006 and 2009, more than 9,400 of the America's twenty-five thousand sewage systems have violated the law by discharging untreated or partly treated human waste, chemicals, and other hazardous materials into rivers, lakes, and elsewhere, according to data from state environmental agencies and the EPA.

---

3      http://geospatial.blog.com/geospatial/2011/02/paying-for-water-.html

## Why Invest in Infrastructure?

Simply put, because we have to. The time is now!

All across America, from small cities to large ones, from California to New York, it is time for cities to start to assess their sewer and water systems and their distribution systems for connectivity and functionality, to assess the roads and bridges for decay and vulnerability, to assess deteriorating gas mains for leaks, to assess drinking-water systems to comply with the EPA standards, to assess waterways, streams, rivers, and lakes for pollutants, illegal dumping, and hazards to the general population.

Questions that cities will need to ask include:

*Is our infrastructure doing what it was designed to do?*

*Has the city outgrown its capacity?*

*Where are the leaks coming from?*

*Why can't we control our flooding problems? Will the dams and levees hold?*

*Can we control flash flooding with a designed comprehensive drainage system?*

*Does our system tie into a much bigger municipality?*

*What about Green space?*

*How do we assess our bridges?*

*What about transportation?*

*Who will fix it?*

*Will the federal government help?*

These are all legitimate questions Congress, mayors, city councils, public works agencies, and taxpayers should be asking to fix our infrastructure.

We know that each city's infrastructure problem is unique and different from another, and what works for one city may not work for another, but the need to assess remains a reality.

In Los Angeles, the infrastructure problems are all-encompassing ranging from water-main ruptures, sewer spills, street and bridge erosion, side-walks, power grid failures, gas line piping, and deferred freeway mainte-nance and flooding.

In Brooklyn looms the Gowanus Expressway, part of the urban interstate building boom of the 1950s. The writer Sarah Goodyear states that no one likes to look at it, and there has been endless talk about what to do with the Gowanus Expressway.[4]

One bright spot in the article is a video clip from the American Society of Landscape Architects that looks at some creative possibilities for repairing parts of cities that have been designed around transportation systems "to the detriment of their own people."

Across the state of New York, wastewater treatment systems are failing and municipalities do not have the funds to adequately repair and replace necessary infrastructure.[5] There are over 600wastewater treatment facili-ties that serve 1,610 municipalities. The facilities range in size from New York City's vast system that processes 1.3 billion gallons of wastewater a day through fourteen facilities to small village systems that process less than ten thousand gallons a day. These facilities provide wastewater treat-ment for more than fifteen million people across the state. The conservative

---

4    http://grist.org/infrastructure/2011-06-15-repairing-our-broken-cities-by-transforming-infrastructure

5    http://www.dec.ny.gov/chemical/42383.html

cost estimate of repairing, replacing, and updating New York's municipal wastewater infrastructure is $36.2 billion over the next twenty years.

The American Water Works Association (AWWA) reports that aging infrastructure is the number one problem.[6] At its 131st Annual Conference and Exposition–ACE (June 10–14, 2012) in Dallas, Texas, the main focus was seeking ways to confront and solve America's water infrastructure challenges, related energy costs and the future of safe water. The key findings in the article include: 1) issues that drive investment or cost are the top concerns among water utility leaders; aging infrastructure is the most pressing concern; 2) more than 75 percent of respondents to a survey have taken measures to reduce energy consumption; 3) more than half of the respondents are implementing asset-management programs; 4) 85 percent of respondents said average water consumers have little to no understanding of the gap between rates paid and the cost of providing water and wastewater services; 5) and nearly half of utility leaders believe that customers will probably be willing to pay higher rates needed to pay for capital improvements.

The American Water Works Association in 2009 went before the House Subcommittee on Water Resources and Environment to urge Congress to create a federal water infrastructure bank to help America invest in its aging water systems, and to stress that the United States is best served by water systems that sustain themselves through consumer rates and other local financing and that the federal government can help by providing access to low-interest loans.

The organization continued by saying the federal water infrastructure bank would be authorized to borrow money through the federal treasury

---

6        http://completewatersystems.com/2012/06/awwa-report-says-aging-infrastructure-is-1-water-problem/

system at very low rates, just as commercial banks do. In turn, the water infrastructure bank would:

- Make low-interest loans for larger water projects, and

- Help those states that want to leverage their SRF capitalization grants, making even more capital available for low-interest SRF loans, noting that a federal water infrastructure bank would require no new taxes.

An article[7] on sustainable business.com states that the United States has the biggest "water footprint" in the world, using nearly 656,000 gallons per capita annually, according to a new report by the Urban Land Institute and Ernst & Young. This amount surpasses usage in China, which uses less than 186,000 gallons per capita annually.

The report, "Infrastructure 2010: An Investment Imperative," found that more and more urban areas throughout the United States—in both dry and rainy locales—are facing growing pressures on their water-infrastructure systems, necessitating both greater investments for overhaul and a change in development patterns that are more conducive to conservation, saying, "most districts do not charge ratepayers full outlays for constructing and maintaining systems… As a result, businesses and households tend to use water inefficiently and don't conserve, even though per-capita water demand could outstrip future availability in some parts of the country."

Furthermore, the article quotes Maureen McAvey, executive vice president of ULI, stating, "Changing growth patterns in response to dwindling resources will not come easy to a nation that is not accustomed to conserving water or land.", "But it is clear that regional and local problems with both water quantity and quality will continue without a broad-based cutback in

---

7    http://www.sustainablebusiness.com/index.cfm/go/news.display/id/20145

public water consumption and a change in how and where we build. Water infrastructure must be viewed through the lens of sustainable growth."

There are solutions though—the report offers solutions to the nation's water infrastructure problem that aim to foster collaboration among different government entities, incorporate land-use planning into infrastructure planning, and accept higher user costs as a necessity. Among the "fixes" specific to water:

- Use federal allocations to encourage the creation of long-range regional management programs to integrate water supply and conservation strategies with population projections, agricultural needs, and utility demand.

- Face reality in that consumers and businesses will have to pay more to ensure reliability and safe supplies.

- Give top priority to repairing and upgrading existing systems.

- Incorporate land use into water management, including restricting development in areas without ample future water resources; using only native species in landscaping; building more compactly to reduce runoff and enhance retention.

- Protect ecosystems to enable more natural storage and restoration.

- Use all available resources, including capturing rainwater, recycling wastewater, recharging groundwater, and making nonpotable water potable.

- Invest in desalinization[8] technology.

In previous annual overview series that analyze the infrastructure needs and compare the infrastructure policies of the United States with those in

---

8    The process of removing the salt from water to make it drinkable.

other countries, the United States continues to lag behind Asia and Europe in investments in transit systems, making its urban areas less competitive globally. In addition to a transportation update, the report includes a look at water infrastructure accessibility and availability, treatment and delivery, and highlights water issues in fourteen US cities as illustrative of the problems looming throughout much of urban America.

The cities are Atlanta, Boston, Chicago, Denver, Houston, Los Angeles, Miami, Minneapolis/St. Paul, New York City, Philadelphia, Phoenix, San Francisco, Seattle, and Washington, DC. Together they are expected to gain an additional sixty million residents between now and 2030, reinforcing the critical need to better coordinate land-use planning with water availability.

Lastly in the article, the writer states that what is most troubling in the research over the years is that the world is moving ahead in rebuilding and expanding its infrastructure without the United States, stating "bottom line the U.S is seriously threatening not only its quality of life now and for the future but also its very basic ability to compete economically with the rest of the world."

In an abstract paper written by Bent Flyvbjerg[9] and titled "Survival of the unfittest: why the worst infrastructure gets built—and what we can do about it," he writes that heads of state, led by US president Barack Obama and China's premier Wen Jiabao, have singled out investments in infrastructure as a key means to create jobs and keep the economy from slumping. China was the first mover when its State Council, in November 2008, passed a $586 billion stimulus plan, mainly investing in infrastructure. In February 2009, the United States followed suit, with Congress passing President Obama's $787 billion New Deal.

---

9    *Oxford Review of Economic Policy*, Vol. 25, Number 3, 2009, pp. 344-367

In an article[10] from the Competitive Enterprise Institute (Issue Analysis, 2012 No.4) by Bonner R. Cohen titled "Fixing America's Crumbling Underground Water Infrastructure: Competitive Bidding Offers a Way Out, the author says America's population is expected to grow by one hundred million—a 30 percent increase—by the middle of the twenty-first century. This growth will put enormous strains on the nation's infrastructure, including roads, bridges, tunnels, and air-traffic control systems. Bonner writes that many of America's vast underground water networks have reached a state of deterioration that exceeds that of the transportation infrastructure above ground and over the next twenty years, upgrading the nation's water and wastewater system is expected to cost between three billion and five trillion dollars. Building and replacing water and sewer lines alone will cost some $660 billion to $1.1 trillion over the same period.

The task at hand is to address the problems besetting those underground networks in the most efficient, cost-effective manner possible at a time when modernizing the nation's water infrastructure is absolutely essential and when governments at all levels—federal, state, and local—are facing substantial budget shortfalls.

Bonner continues by stating that inserting some market discipline into the process would go a long way toward achieving that goal. Opening up the bidding process under the principle of "may the best technology win" will immeasurably improve the quality of America's underground water infrastructure in a cost-effective fashion. Competitive bidding can serve as an essential safeguard against the influence of politically preferred providers of government services (end).

As all of America's sewer and water infrastructures have passed the one hundred-year mark, the materials used then have far outlived their life

---

10    http://www.scribd.com/doc/88877493/Bonner-Cohen-Fixing-America-s-Water-Infrastructure

expectancy. New technology and new materials are replacing aging manholes, bridges, roads, buildings, structures, pipes, electrical systems, and cables, along with how to repair and replace them.

Urban sprawl, immigration, and the general population have put a strain on America's infrastructure and the need for repair and upgrading is long overdue.

As record yearlong floods continue, as heavy snows fall, as bridges rust and weaken, as roads give way to collapse, and as pipes continue to collapse and manholes surcharge, we are reminded of the need to fix our infrastructure.

We need not look at these repairs as costly maintenance items put on capital-improvement budgets but as returns on investments for our children and their children.

In another one hundred years, cities will be faced with the same problems we face today, and they should have better and more advanced technology based on the foundation we lay today.

Today we are being outspent by other countries in investing in our infrastructure. How can we as a country think of ourselves as leaders when we fall behind in investing for the future?

## Transportation Projects Across the United States

Ryan Holeywell and Daniel Lippma, writing for *Governing*[11] magazine, discuss the "5 Biggest U.S. Infrastructure Projects, Plus 5 at Risk" across the United States from Washington, DC, to California, citing these ten projects as the nation's highest-profiled infrastructure projects—five making steady progress, and five facing serious challenges—under way right now.

1) Dulles Transit Extension, Washington, DC, (the biggest expansion in the history of the Metro system, at a cost of $6.2 billion).

2) Otay Mesa East, San Diego (a third land port adding to the existing two border crossings to alleviate bottleneck traffic, at a cost of $715 million).

3) O'Hare Modernization, Chicago (a modernization effort to increase the faciliity's efficiency and capacity, at a cost of $8.8 billion)

4) Cresent Corridor Expansion (a freight rail network that runs through thirteen states and connects New Orleans to New Jersey by the railroad company Norfolk Southern, at a cost of $2.5 billion).

5) Alaska Way Viaduct, Seattle (a combination of viaduct, roadway, and tunnels, at a cost of $3.1 billion).

---

11    http://www.governing.com/topics/transportation-infrastructure/gov-5-biggest-us-infrastructure-projects-plus-5-at-risk.html

The other five in trouble are:

6) Columbia River Crossing, Oregon/Washington (a joint venture replacing the existing Columbia River bridge that connects Portland to the suburbs of Vancouver, Washington, at a cost of $3.5 billion).

7) Denver FasTracks, Denver (122 miles of commuter and light rail and eighteen miles of bus rapid transit service across the Denver area, at a cost of $7.8 billion)

8) NexGen, (an FAA project to revolutionize air travel in the country by switching from radar-based to satellite-based flight-tracking technology, along with other technological advances like improved weather forecasting and communications, estimated between twenty and twenty-seven billion dollars). The Government Accountability Office (GAO) says the cost could get as high as $160 billion.

9) California High-Speed Rail, Los Angles and San Francisco (800 miles of high-speed rail lines connecting the two cities. The train would reduce air pollution and ease congestion on the state's clogged freeways, at a cost upward from forty-three billion to at least ninety-eight billion dollars).

10) Second Avenue Subway, Manhattan (Phase 1: a two- mile section with two new tunnels and three new stations on the Upper East Side by MTA, at a cost of $4.45 billion; Phase 2: 8.5 miles extending the line uptown to Harlem and downtown to the financial district, at an expected additional cost of thirteen billion dollars).

Other cities that have made infrastructure work a priority include:

1.  Baltimore, where city officials, business owners, neighborhood leaders, and residents have formed a coalition and launched the Healthy Harbor Initiative, a plan to make the Inner Harbor fishable and swimmable by 2020. The Baltimore Waterfront Partnership is launching the community-wide effort.[12]

2.  Boston is fixing its bridge infrastructure[13] using "accelerated bridge construction" techniques, a collection of technologies and methods that can shave months if not years off the process of building and replacing critical infrastructure and Massachusetts is at the forefront of a national effort aimed at putting drivers first. Additionally, commission members are deliberating over what they are calling the Blue Act—the blue symbolizes water, they said, and is in sync, from a color perspective, with terms like brownfields and green energy. That fund would be intended to help cities and towns address infrastructure costs and ensure safe and sustainable water supplies.

3.  Chicago is investing seven billion dollars to fix its crumbling water pipes and sewer lines. The program, slated to be one of the largest investments in Chicago's history, will touch "nearly every aspect of the city's infrastructure network"—from its largest airport to streets, schools, community colleges, parks, the water system, and the commuter rail stations.[14]

---

12   http://www.bayjournal.com/article/waterfront_partnership_launches_baltimore_harbor_cleanup

13   http://www.nytimes.com/2012/04/18/us/rapid-construction-techniques-transform-infrastructure-repair.html?pagewanted=all

14   http://www.circleofblue.org/waternews/2012/world/chicago-spearheads-7-billion-plan-to-fix-its-crumbling-infrastructure

4. Cincinnati has studied and prepared a map to identify hillsides susceptible to landslides. The subject map is called "Landslide-prone bedrock hillsides within the city of Cincinnati." The US Geological Survey of the Department of the Interior recommended that the city contract a comprehensive survey identifying specific landslide-prone locations and the degree of their susceptibility to sliding as the first step in a long-range program to deal with the slide problem in Cincinnati.

5. Minneapolis replaced the aging and design-flawed Mississippi River Bridge that collapsed over the Mississippi River with the I-35W Saint Anthony Falls Bridge, maintained by the Minnesota Department of Transportation (MDOT). The planning, design, and construction processes were completed faster than normal and the bridge opened ahead of schedule on September 18, 2008. Within hours of the previous bridge's demise, politicians pledged to rebuild the bridge at an accelerated pace, with a cost of eighty-five million dollars for the cleanup and recovery.

6. Philadelphia is tackling its challenges of maintaining and upgrading its infrastructure head-on through the city's "Green City Waters Program[15] (GCCW)." The program seeks to integrate water resources management "into the socioeconomic fabric of the city," states Jeffrey Featherstone, who examined Philadelphia's GCCW program as a case study for the implementation of proper green infrastructure. In a July 2008 article on Stateline.org,[16] staff writer Pamela Prah states that

---

15    http://www.newswise.com/articles/low-impact-green-solutions-fix-older-city-water-infrastructures

16    http://www.pewstates.org/projects/stateline/headlines/govs-turn-to-fixing-infrastructure-85899387219

improving the nation's crumbling bridges, roads, and sewer systems is a $1.6 trillion-dollar problem that governors intend to address, citing Ed Rendell, who said that when president Dwight D. Eisenhower was in the White House more than forty-five years ago, 11.5 percent of nonmilitary federal spending went for infrastructure, compared with less than 2.5 percent today. The federal Highway Trust Fund, which pays for roughly 45 percent of the nation's roads and bridge building, was expected to run out of money in 2009, falling $3.3 billion short of needed transportation funding.

Here in Atlanta, the Clean Water Atlanta Initiative, and the Department of Watershed Management, whose motto is "Protecting Our Future," was spearheaded by the city, Mayor Franklin and her administrations efforts, starting in 2002, and together they orchestrated the greatest single comeback for any US city within the time frame given and today serve as a best-in-class city.

## *Global Infrastructure*[17]

1. China- leapfrogs the rest of the world when it comes to building modern transport infrastructure, investing hundreds of billions of dollars in new roads, dams, mass transit, high-speed rail, ports, and airports. The government has directed most of the $600 billion in stimulus funds to large-scale infrastructure, including nearly ten thousand miles of new high-speed rail to be completed by 2020.

2. The European Union- provides $630 million to member nations to spend on rail links between countries, including high-speed rail lines.

3. France- injected $1.21 billion in stimulus funds into its transportation sector in 2009, and moved ahead with plans to double its high-speed rail system to 2,500 miles by 2020. The government also aims to be the world leader in developing infrastructure to support the use of electric and hybrid electric cars.

4. Germany- a consortium of German industrial, energy, and finance companies pursues a $556 billion solar energy project to transport solar-generated electricity from state-of-the-art plants in the Sahara Desert to Germany and other European

---

17    http://www.sustainablebusiness.com/index.cfm/go/news.display/id/20145

countries. It could supply as much as 15 percent of Europe's energy needs by 2050.

5. India- has a $475 billion plan[18] over five years for modernization of highways ($ 75 billion), development of civil aviation ($ 12 billion), development of irrigation system ($ 18 billion), development of ports ($ 26 billion), development of railways ($ 71 billion), development of the telecom sector ($ 32 billion), and development of power ($ 232 billion).

6. Japan- has employed public works stimulus to boost its lackluster economy for two decades, building new roads and new airports and expanding its "bullet" trains. Now it faces population losses and an aging population, leaving it with an overdeveloped infrastructure system offering more capacity than is warranted by demand.

7. Puerto Rico[19]- the U.S. EPA administration has awarded nearly seventy-two million dollars through the American Recovery and Reinvestment Act of 2009 to help the commonwealth and local governments finance overdue improvements to wastewater and drinking-water systems and conduct water-quality planning essential to protecting human health and the environment.

8. Singapore- enhances its reputation for acclaimed infrastructure with the completion of the Marina Barrage, a $170 million hydroelectric dam project that integrates flood control, green technologies, and recreational features.

---

18    Indiamanufacturingshow.com/infrastructure.html

19    Underground magazine, October 2009, waterworks news

9. South Korea- in the pipeline is a ninety-three-mile under-ground road network in Seoul budgeted at nine billion dollars, a three-billion dollar expansion of Incheon Airport; $2.3 billion in green energy initiatives; and a nineteen-billion dollar clean-up of major rivers, all demonstrating the country's ongoing commitment to advancing its infrastructure systems.

10. The United Kingdom- adopts a combination of large-scale and small-scale infrastructure initiatives to reduce congestion around London. The seventy-three-mile Crossrail tunnel will connect Heathrow Airport to the eastern suburbs; and outside the city, the Eurostar bullet train now extends to Amsterdam.

## Background of the Clean Water Act

Before the turn of the twentieth century, it was common practice in many American cities to discharge household excrement[20] and industrial wastes openly in the streets.

In the late 1870s, Atlanta's sanitation districts were implemented and sewer services were limited to the central business district.[21] In this section, the writer provides some case background on Atlanta's sewer beginnings.

In 1911, when Atlanta's first wastewater treatment plant was built, both storm water and sewage were piped to treatment plants. After rainfall events, overflow pipes emptied into creeks and streams to act as safety valves, releasing a toxic combination of storm water and raw sewage called combined sewer overflows (CSOs). Since the early 1900s, new housing developments in the metropolitan Atlanta area have been served by a separate sewer system in which sewage and storm water are collected in separate pipes, sewage is treated and discharged into a designated receiving stream, and untreated storm water is discharged directly into a receiving stream.

Combined sewer systems collect wastewater and storm water in a single pipe. During dry weather, combined sewer systems transport wastewater directly to a wastewater treatment plant. In heavy rainfall events, the volume of the mixed storm water and wastewater effluent[22] can overwhelm the capacity of a municipality's sewer system or network of wastewater

---

20  Waste matter discharged from the bowels: feces.

21  http://webmit.edu/dusp/dusp_extension_unsec/projections/issues_8/issues_8_jelks.pdf

22  A substance (usually liquid) outgoing or emanating.

treatment plants. When the capacities of the main wastewater treatment plants are exceeded, combined sewer systems overflow.

CSO-contaminated waters contain pathogenic organisms[23] from untreated human, animal, and industrial waste; toxic materials like petroleum products, heavy metals, pesticides, and other organic compounds; and floating trash and debris washed into the sewer system (US EPA 1995).

Raw sewage carries a variety of human bacteria and viruses. Depending on the amount and concentration of the sewage and mode of people's exposure to it, these accompanying bacteria and viruses cause illnesses ranging from hepatitis and gastroenteritis to cholera, skin rashes, and infections like giardiasis (CDC 2002).

## 1. The Federal Water Pollution Control Act of 1948

This was the first major US law to address water pollution.[24] Growing public awareness and concern about controlling water pollution led to the 1972 amendment. Passed and implemented in 1972 by the Environmental Protection Agency (EPA), its focus was at the time on states and Indian tribes that polluted rivers and streams with contaminants of a chemical nature. The amendments to the 1972 act have given it its current shape. The focus was also on implementing an act that focused on regulating discharges from facilities such as sewage plants and industrial facilities. No attention was being paid to runoff into storm drains, the streets, construction sites, farms, or other "wet weather sources." Over the decades, a shift in focus by the EPA is being addressed by a more "holistic" watershed-based strategy. Emphasis now is being placed on protecting healthy waters and restoring impaired ones.

---

23    Bacteria that causes infection.

24    http://www.epa.gov/lawsregs/laws/cwahistory.html

## 2. The Clean Water Act

This is a federal law that regulates the discharge of pollutants into the nation's surface waters, including lakes, streams, wetlands, and coastal areas.

The Clean Water Act is administered by the EPA, which sets water-quality standards, handles enforcement, and helps state and local governments develop their own pollution-control plans.

The original goal of the Clean Water Act was to eliminate the discharge of untreated (not treated with a reagent, dye, or chemicals) wastewater from municipal and industrial sources, making waterways safe for swimming, fishing, and boating.

The Clean Water Act also requires businesses to apply for federal permits to discharge pollutants into waterways, as well as to reduce the amount of their discharges over time.[25]

The Clean Water Act consists of two major parts, one being the provisions which authorize federal financial assistance for the construction of municipal sewage treatment plants. The other is the regulatory requirements that apply to industrial and municipal dischargers.

Under the Clean Water Act, the EPA sets national water-quality criteria and specifies levels of various chemical pollutants that are allowable under these criteria. The discharge of regulated chemicals into surface waters is controlled by the National Pollutant Discharge Elimination System (NPDES), which requires polluters to obtain federal permits for every chemical they discharge.

The 1977 amendments established the basic structure for regulating the discharge of pollutants into the waters of the United States. The amendments:

25    www.answers.com

- Gave the EPA the authority to implement pollution-control programs such as setting wastewater standards for industry.

- Maintained existing requirements to set water-quality standards for all contaminants in surface waters.

- Made it unlawful for any person to discharge any pollutant from a point source[26] into navigable waters, unless a permit was obtained under its provision.

- Funded the construction of sewage-treatment plants under the construction-grants program.

- Recognized the need for planning to address the critical problems posed by nonpoint source[27] pollution.

Revisions in 1981 streamlined the municipal construction-grant process, improving the capabilities of treatment plants built under the program. Changes in 1987 phased out the construction-grant program. Over the years, many other laws have changed parts of the Clean Water Act.

The Clean Water Act is the primary federal law in the United States governing water pollution.[28] Commonly abbreviated as CWA, the act established the goals of eliminating releases of high amounts of toxic substances into water, eliminating additional water pollution by 1985, and ensuring that surface waters would meet standards necessary for human sports and recreation by 1983.

In the Water Quality Act of 1987, Congress responded to the storm-water problem by requiring that industrial storm-water dischargers and municipal governments separate storm sewer and obtain NPDES[29] permits by

---

26    Waste discharge from discrete sources such as pipes, and outfalls.

27    Storm water runoff from farm lands, forests, construction sites, and urban areas.

28    http:en.wikipedia.org/wiki/Clean_Water_Act

29    National Pollution Discharge Elimination System

specific deadlines. To combat nonpoint-source pollution, the EPA initiated numerous programs and grants to aid the public in improving local water quality.

Prior to 1987, programs were primarily directed at point sources (waste discharge from industrial facilities, including manufacturing, mining, oil and gas extraction, service utilities, military bases, municipal governments, some agricultural facilities such as animal feedlots).

*Note*:

Since the early 1980s, the City of Atlanta had faced challenges in complying with increasingly stringent federal Clean Water Act standards. Those challenges eventually led the city to entering into two Consent Decrees (CDs) to address the operation of its wastewater facilities and combined sanitary sewer overflows (CSOs and SSOs) within the city.

## EPA Involvement

The Environmental Protection Agency (EPA) is an agency of the US government that sets and enforces national pollution-control standards[30].

This agency was established in 1970 by President Richard Nixon to supersede a swelter of confusing and ineffective state environmental laws by consolidating fifteen components from five agencies for the purpose of grouping all environmental regulatory activities under a single agency.

The purpose of the EPA is to ensure that all Americans and the environment in which we live are safe from health hazards.

The EPA has a number of goals: clean air, clean and safe water, safe food, preventing and reducing pollution, water management and restoration, and oversight of the cleanup of abandoned waste sites.

The EPA established a handbook in 1991 for all states nationwide with the purpose of using these federal standards as a guide for Sewer System Infrastructure Analysis & Rehabilitation work. The role the EPA plays in mandating that Atlanta repair its infrastructure comes in conjunction with the Water Pollution Control Act (1972) Regulatory Requirements, providing that the EPA evaluate the conditions of cities' sewer systems for the purpose of investigating and documenting sewer-system discharge, infiltration, and exfiltration of the system through Analysis and Sewer System Evaluation Survey (SSES). In addition, cities had to indicate the most effective means/methods of rehabilitation to correct the sewer pipes and manholes. Cities had to estimate flow data at treatment plants and determine the causes of industrial waste either legally or illegally into the system. Cities had to come up with "best practices" to determine the method(s) and costs of rehabilitation and the most efficient method(s) to repair the system.

Section 204 of the 1981 Amendments to the Clean Water Act requires that applicants for EPA grants, after applying for funding, must after one year of operation verify to the EPA that the project meets design specifications and effluent limitations. The purpose of the certification requirements is to ensure that effective sewer-rehabilitation projects were carried out through the grant assistance program.

In or around the late 1870s, Atlanta implemented sewer services limited to the central business districts in downtown.

Atlanta has owned and operated a drinking water and wastewater system for more than 130 years.[31] Over the years, the city has invested several billion dollars in the infrastructure that supports those systems. The systems have undergone many expansions and improvements since their creation in the late 1800s. Atlanta spent more than one billion dollars making improvements to its sewer system during the 1990s. Under the Consent Decree negotiated in 1998 and 1999 with the EPA, the 2002 Capital Improvement Program (CIP) calls for a $3.9 billion investment in drinking water and sanitary sewer and wastewater treatment improvements. Atlanta developed an asset-management program to ensure that the investment being made today in its water and wastewater systems will benefit Atlanta and its downstream neighbors for decades into the future. Atlanta will also use asset-management tools to make sure that its watershed infrastructure is maintained in such a manner that it protects the public health and the environment.

---

31    City of Atlanta, Dept. of Watershed Management (Asset Management Case Study)

*Notes:* Mayor Bill Campbell proposed a plan worth $150 million to make repairs to Atlanta's infrastructure, but he was halted by the city council as the city prepared to host the 1996 Olympic Games.

## Upper Chattahoochee Riverkeeper Involvement

Established in 1994 and devoted to the protection and preservation of the Chattahoochee River system from the North Georgia mountains to the Florida border, the organization is dedicated solely to protecting and restoring the Chattahoochee River basin and its drinking water for millions of people. The Chattahoochee River is the most heavily used water source in Georgia.[32] More than 150 municipal and industrial wastewater treatment facilities are permitted to discharge into the upper Chattahoochee River basin between the town of Helen and the West Point Dam. The quality of the water is a result of the complex interaction of natural and human influences on land and water. Metropolitan Atlanta, the largest and fastest-growing metro area in the Southeast, is in the river basin's headwaters.

The City of Atlanta withdraws more than 300 million gallons per day and discharges more than 200 million gallons[33] of wastewater while the river keeps flowing toward the sea. A seventy-mile stretch below the city is arguably among the most polluted sections of any river in the nation, but still the river endures. It fills up three more major reservoirs and provides hydroelectric power, commercial navigation, and flood control before finally being set free to feed one of the world's most productive estuaries.

From its birth in the mountains to its surge into the sea more than 500 miles later, the Chattahoochee is the quintessential American river, complete with all the paradoxes that designation entails. It is used and abused,

---

32      www.ucriverkeeper.org

33      http://www.sherpaguides.com/georgia/chattahoochee/natural_history

coveted and ignored, and loved and cursed, as it provides drinking water for more than half of all Georgians and recreation for more than twenty-five million Americans.

Since prehistory, the Chattahoochee has been a working river. Cherokees, Creeks, and tribes long before them depended on it for drinking water, food, and transportation. Never, though, has it worked harder, as the river's basin today is the smallest in the nation, serving as a primary source of drinking water for a large metropolitan area.

The Upper Chattahoochee Riverkeeper uses strategic legal actions to stop pollution and promote policies that are protective of the river and its watershed[34].

In an article[35] by Sally Bethea of Upper Chattahoochee Riverkeeper, she explains it this way; In the late 1990s the Upper Chattahoochee Riverkeeper, the Upper Chattahoochee Riverkeeper Fund, the EPA/EPD, the state of Georgia at the request of the Georgia Department of Natural Resources, and a concerned citizen residing along the river in Newnan, Georgia, filed a lawsuit in a federal court in Atlanta alleging the city was in violation of the Clean Water Act. Citing the original Clean Water Act, the plaintiffs stated under 33 U.S.C §1251 et. seq because the city holds the permit of the National Pollution Discharge Elimination System (NPDES), that made the city responsible for managing, operating, and maintaining the wastewater collection system and treatment facilities. The Clean Water Act made it unlawful to discharge any pollutant from a point source into navigable waters, unless a permit was obtained.

---

34    A basin-like landform defined by high points and ridgelines that descend into lower elevations and streams and valleys.

35    http://www.waterkeeper.org/ht/a/GetDocumentAction/i/9998, Sally Bethea

In 1995, Atlanta was sued by the Upper Chattahoochee Riverkeeper[36] for failing to comply with the Clean Water Act. The Upper Chattahoochee Riverkeeper filed a lawsuit in federal court against the City of Atlanta for failing to control the discharge of raw sewage and other pollutants into these waterways from combined sewer overflows (CSOs).[37] The UCR won the case in 1997 and settled in 1998.

The lawsuit ultimately led to two federal consent decrees that mandate CSO compliance with the Clean Water Act in 2007 and SSO compliance by 2014.

The Combined Sewer Overflow (CSO) facilities in the areas shown— Proctor Creek/ North Avenue, Greensferry, Tanyard Creek, McDaniel Street, Custer Avenue, Intrenchment Creek, Clear Creek, which control, treat, and maintain wastewater discharge—were not properly containing raw sewage and untreated wastewater that found its way into the Chattahoochee and South rivers and their tributaries. Further, the lawsuit contends the city was in violation of NPDES permits issued by the EPD with respect to the wastewater-treatment facilities R.M. Clayton, Utoy Creek, and South River.

Combined sewer/storm-water systems and the relatively low capacity of the Chattahoochee River greatly impact the city's wastewater and storm-water situation. Over time, the city had made limited capital investment, resulting in an aging distribution system that has required larger-than-normal investment in pipe, valve, and meter infrastructure.

---

36    City of Atlanta, Dept. of Watershed Management (Asset Management Case Study), Federal Consent Decree

37    http://www.ucriverkeeper.org/enforcement-hightlights.php

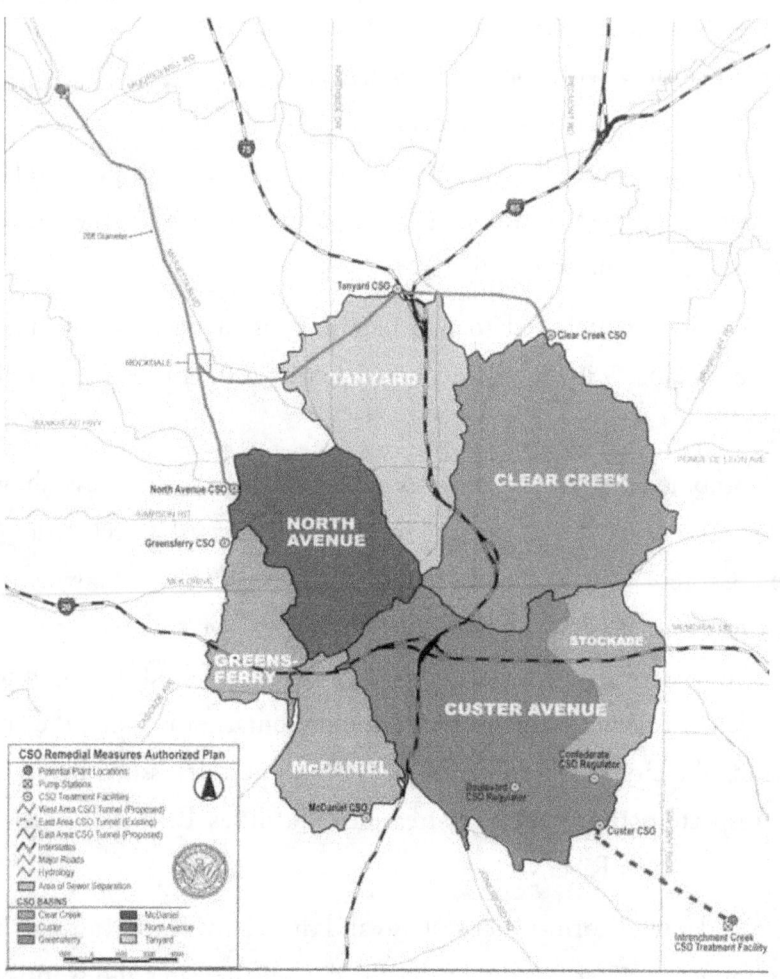

Lastly, the city was in violation of the Clean Water Act by discharging pollutants from unpermitted point sources. The plaintiffs were seeking injunctive relief and penalties only for those violations relating to the CSO facilities, and an amendment to the original consent decree for work to be performed with respect to the said treatment facilities.

In another powerful article[38] written by Sally Bethea and titled "Cleaning Up Atlanta: Sewage Overhaul," she writes about the background history of

---

Atlanta's sewers and how it led to the Upper Chattahoochee Riverkeeper's involvement in bringing Atlanta into compliance with the Clean Water Act.

She writes that in 1934, construction began on a new sewer system, financed by the federal Works Progress Administration. Prior to that time, half of all sewage was simply dumped into streams leading into the Chattahoochee River. With each new home, hotel and high-rise office building, the number of sewer connections to the aging system increased. By the 1970s, the city's sewer system was so overloaded when it rained that it discharged raw sewage directly into city creeks, leaving toilet paper hanging in trees and human waste rotting in stagnant pools. Environmental officials at all levels—city, state, and federal—knew about the situation, which threatened public health, recreational areas, and property values, but they did nothing. Because of the combined system, rain that flowed into storm drains was funneled into the same pipes that carried household and industrial sewage to treatment plants. During torrential downpours, the sudden inflow of rainwater swamped the sewage treatment plant. The resulting overflow of storm water and untreated sewage flowed into creeks and rivers, carrying waste and other matter downstream. In the rest of the city, sanitary sewer overflows (SSOs) polluted neighborhood streams even during dry weather—thanks to decades of failure to maintain, repair, and replace 1,800 miles of sewer lines. The city's three sewage-treatment plants were inadequate and ill-maintained.

Sally Bethea goes on to write that in 1989, nearly four million tons of phosphorus flowed down the Chattahoochee River and West Point Lake (65 miles downstream of Atlanta) every year. In the fall of 1989, the city's R.M. Clayton sewage treatment plant, the largest such facility in the Southeast, dumped 200 million gallons of raw sewage into the river during a storm event. The article continues that massive spills such as this one were routine during the 1980s and 1990s. Irate citizens took their concerns to the state capitol and demanded that something be done to protect their lake and the river; they called Atlanta's sewer-disposal system an "abomination" and a "disgrace".

To tout some of Upper Chattahoochee Riverkeeper's 2001 accomplishments, they:

- Organized eleven stream and river cleanups and conducted 145 speaking engagements to schools, civic and business groups, neighborhood associations, legislators, and other conservation groups.

- Helped secure a new Interim Instream Flow Policy for Georgia that was adopted by the DNR board in May 2001.

- Monitored the progress of the City of Atlanta in meeting its responsibilities under the federal consent decree negotiated to resolve URC's lawsuit against the city for violations of the Clean Water Act at its sewage facilities.

- Continued the daily bacteria monitoring of the river within the National Recreation Area in partnership with the US Geological Survey and the Park Service.

*Urbanization*

## 1. History of Atlanta

Atlanta was founded in 1837, when Georgia decided to build a railroad to the Midwest, a century after Savannah, the state's oldest city.[39] The three dominant forces affecting Atlanta's history and development have been transportation, race relations, and the Atlanta spirit. At each stage in the city's development, these three elements have come into play. In 1837, engineers for the Western and Atlantic Railroad (a state sponsored project) staked out a point on a ridge about seven miles east of the Chattahoochee River as the end of a rail line they planned to build south from Chattanooga, Tennessee. The stake marking the founding of "Terminus" was driven in the ground in 1837. The town that emerged around this zero milepost was Terminus, which literally means "end of the line." In 1839, homes and stores were built there and settlements grew. Between 1845 and 1854, rail lines arrived from four different directions, and the rapidly growing town quickly became the hub for the entire southern United States.

Atlanta began as a settlement located at the intersection of two railroad lines, and it was incorporated in 1845. It was the railroad that brought Atlanta into being and eventually connected the rest of the state and region.

By 1846, however, two other railroad lines had converged with the Western and Atlantic in the center of town, connecting it to as far-flung areas of the Southeast and boosting the city's growth.

---

39    http://www.geogiaencyclopedia.org/nge/Article.jsp?id=h-2207

During the American Civil War, Atlanta as a distribution hub became the target of a major Union campaign, and in 1864 Union General Sherman's troops set on fire and destroyed the city's assets and buildings, saving churches and hospitals.[40] After the war, the population grew rapidly, as did manufacturing, while the city retained its role as a rail hub. Coca-Cola was launched in 1889, and the city added new "streetcar suburbs."

From the time of its founding in 1837, Atlanta took 122 years to reach the one million mark in population.[41] Ceremonial shovels from numerous groundbreaking events lined a wall in City Hall in the 1960s; they were so many that the then mayor of Atlanta Ivan Allen Jr. quipped, "We may have done too good a job."

The essential question is: How does an area as big and vibrant as Atlanta retrofit itself so that it can accommodate growth with its existing systems, roads, buildings, culture, and infrastructure?

The population in and around Atlanta has swelled beyond five million people. That's enormous for an area that started as a train station in the forest.

In 1925, Ivan Allen Sr. and W.R.C. Smith of the Atlanta Chamber of Commerce launched a national advertising campaign entitled "Forward Atlanta," which was designed to lure new businesses to the city and to encourage national corporations to establish their regional headquarters there. The campaign was extremely successful, bringing thousands of jobs and adding an estimated thirty-four million dollars in annual payrolls to the city's economy.

Growth has brought its challenges. For the people who work and play in Atlanta, the area is home to the nation's third-longest commute time. That daily grind on the roadway symbolizes our deteriorating air quality and our

---

40    http://en.wikipedia.org/History_of_Atlanta
41    http://www.sustainableatlanta.org/report/Sustainability%20Report.pdf

increasing urban sprawl. Carbon dioxide emissions in Georgia grew over 30 percent from 1990 to 2005. Gains in population and commerce also remind us that our area relies on the smallest water supply of any major metropolitan area in the country, a condition amplified by one of the worst droughts in nearly a century.

Today, Atlanta is a major business city and the primary transportation hub of the southeastern United States (via highway, railroad, and air).

In my view, in the early to mid 1990s, Atlanta saw a rise in population after the 1996 Olympics. The city started experiencing a high growth record. People from all over the world came to see what the South had to offer. Now the world has come to know why it is called "Hotlanta" (referring to the southern climate, and the city's recent prosperity). Having said that, the climate and culture of Atlanta were very appealing and people decided not to go back home and stayed here. Others, after visiting a few months or a couple of years, decided to move south to Atlanta.

This influx of people has been steady since around 1990 and has led to the city's growing housing development, business development, swelling highways, and the growth demand of an already overwhelmed combined sewer system and drinking-water system. Construction started popping up all over midtown and downtown in preparation for the Olympics. Condominiums were infilling every vacant lot. The Olympic Village was being constructed at Georgia Tech's campus, the Olympic Stadium down-town, and the Soccer Stadium at Morris Brown College. New hotels were under construction or being renovated, MARTA transit was building a new CNG bus facility to house its new fleet of CNG buses at Perry Boulevard, and Piedmont Park was on the design table for a drainage makeover.

All these sites included new manholes, sewer piping, storm piping, and catch basins. All the surrounding pipes had to be upsized to accommo-date the volume of these added structures onto the system that had not yet started to be overhauled. The added strain on sewer pipes that are too small to handle the volume of waste flow caused the sewage not to flow as fast, getting to the manholes and pipes downstream, heading to the treatment plant, and causing backup. This added pressure in the pipes caused already weak and damaged pipes to crack, break, and fail. When pipes get to this stage, they are very vulnerable to sagging, dislocating, and separation. When this happens, the waste has nowhere to go but down into the soil, causing soil contamination or surging out of the manhole.

## 2. Population Growth and Usage Reduction[42] in Atlanta

From 2000 to 2008, Atlanta experienced unprecedented population growth, adding almost 30 percent to its population. But it has done so with an emphasis on proper resource management—smart-growth policies, in-fill housing instead of sprawl, extensive capital investment in its systems, a diligent leak-detection and repair program, and conservation. In fact, Clean Water Atlanta served as a launching pad for green initiatives like the construction of a green roof at City Hall, land acquisition for parks, energy conservation projects, and a Green Building Ordinance.

Atlanta's population and demographics[43]:

- 5,475,213 (2009 est.) in the twenty-eight-county Atlanta Metropolitan Statistical Area, designated by the Metro Atlanta Chamber of Commerce[44]

- 4,124,300 in the ten-county Atlanta region

- 540,922 (2009 est.) in the City of Atlanta

---

42    City of Atlanta Online, http://www.atlantaga.gov/media/nr_cleanwater_080509.aspx

43    http://www.atlanta.net/visitors/population.html

44    Atlanta Regional Commission, 2011; Census Bureau 2009

In the past six years, the metro area has added more than 458,568 people annually, more than any other metro area in the United States. At this growth rate, the projected population of Atlanta's twenty core counties for 2020 is 6.4 million.

Growth means the Atlanta population has become more diverse. In fact, the city's racial diversity is greater than that of the nation as a whole. Atlanta is also younger than the US population, with an average age of thirty-four compared to thirty-six for the population as a whole. More than 35 percent of those who moved here from 2000 to 2004 came to Atlanta from a different state. The largest percentage of people came from New York and New Jersey.

According to the 2010 US census,[45] the population of Atlanta was 420,003, although the metropolitan area (comprising twenty-eight counties and more than six thousand square miles) has a population of more than 5,268,860 million people and is the ninth largest city in the United States. It is also one of the most important commercial, financial, and transportation centers of the southeastern United States.

In the City of Atlanta Business Model (part of the Atlanta Case Study Project),[46] Harvey K. Newman and Sanchita Sarkar report that Atlanta is a municipality that contains only 10 percent of the residents of a metropolitan area of more than five million. This means the city's daytime population swells with commuters, and visitors require services from Atlanta while contributing little in revenue. In 2002, the US Census Bureau estimated that Atlanta had the largest percentage increase (62.4 percent) and the largest absolute number of people coming into the city on a daily basis (259,957) of any city of its size in the nation. These "day timers" make

---

45    http://www.geogiaencyclopedia.org/nge/Article.jsp?id=h-2207

46    http://fiscalresearch.gsu.edu/atlanta_case_study/Business%20Model%20Case%Study%report.pdf

911 emergency calls, cause wear and tear on roads and bridges, and require many of the same routine services from the city government as residents.

## 3. A Change From the Normal in Community Development to "Go Green"

Novare Group, a successful Atlanta real estate developer, purchased the Glenwood Park land in 2000 and created a mixed-use plan for the land that featured a large office component sharing parking with condominiums, and a grocery-anchored shopping center. The land was rezoned to allow the initial mixed-use development, but given the economic realities of 2001, the development never made it.

Atlanta's development to address urban sprawl, smart growth, community connectivity, green space and neighborhood patterns and designs that emphasize the creation of vibrant, equitable communities that are healthy, walkable, mixed-use, and a break away from conventional development was realized in 2001, when Katharine Kelly, Walter Brown, and Charles Brewer formed Green Street Properties. They were invited by the Novare Group, which had purchased the site in 2000, to invest in the Glenwood Park project and take over the development.

### a) Residential: Glenwood Park

In an article[47] entitled "Glenwood Park, an UnSprawl Case Study," writer Simmons B. Buntin explains how this redevelopment would reshape conventional thinking on how communities in the future should look, how they will be financed, and how they incorporate existing neighbors.

---

47      http://www.terrain.org/unsprawl/17

Glenwood Park was a brownfield[48] redevelopment (an industrial site that had mostly been used as a concrete recycling facility) in an infill located two miles from the center of downtown Atlanta near Interstate 20, a mile from two different MARTA rail stops, and directly on an active bus route that leads to downtown.

The vision of Green Street Properties for the site was clear from the start, as stated in the article; Glenwood Park would be a real neighborhood that featured a traditional fine-grained mix of different housing types, as well as retail activity, office space, civic buildings, and recreational assets. Green Street properties states making Glenwood Park as "green" as possible, with the dedication to environmental design, would be clearly manifested in the brownfield redevelopment dedicated to urbanism by the planners.

The design team pursued Planned Development Mixed Use (PDMU) zoning. All of the project's financing for the land-development work had been provided by a small group of "insiders." No bank debt was used. As a result, the developer was able to "quickly make a decision" about the development process without any outside financial pressure.

Glenwood Park would emphasize the public realm—community, diversity, the quality and character of streets, sidewalks, parks, plazas, and other public spaces. The commercial center would have retail establishments that serve the practical everyday needs of both Glenwood Park and its surrounding neighborhoods—needs that were not being met. Glenwood Park would be walkable, with plenty of interesting things to walk to. Sidewalk and street designs would emphasize pedestrian comfort and safety. Glenwood Park would be designed to allow a great deal of flexibility in how the neighborhood would evolve over time, and meet pedestrian access to basic services for community connectivity.

---

48    An abandoned or unused piece of industrial or commercial property underutilized or contaminated that, when cleaned up, can be reused.

The results:

All homes would meet Atlanta's EarthCraft House program standards. Glenwood Park is one of only five projects serving as a pilot for the EarthCraft Communities program. The community would also be selected as the site of a 2005 Southern Living Idea House, featured in the August 2005 issue of *Southern Living* magazine.

The project is relatively high density, providing residents the opportunity to drive less. By one estimate, Glenwood Park would save 1.6 million miles of driving per year over what residents would have to drive if they instead lived in a "typical" new Atlanta development.

This type of development would lay the foundation for reshaping how residential communities live, work, and play.

*Note:* Urban sprawl causes many office workers to have longer commute times.

## b)  Commercial: Technology Square

Technology Square, commonly called Tech Square,[49] is now an extension of the Georgia Tech downtown campus. An area that was previously vacant surface parking lots would come to symbolize campus connectivity in Atlanta's downtown district.

Technology Square is a mixed-use district on the block of Fifth Street between the Downtown Connector and Spring Street. This reconstruction/revitalization in 2007 of the Midtown neighborhood for Georgia Tech includes several academic buildings and provides access to the campus via the Fifth Street Pedestrian Plaza Bridge. It also contains restaurants, retail shops, condominiums, office buildings, and a hotel.

---

49    http://www.triposo.com/poi/W__42706836

*Note:*

This type of development would showcase commercial life, work, and study.

## Atlanta's Consent Decree

The following parts are excerpts from the original Consent Decree[50] that represented the issues brought forth on the City of Atlanta as filed in the United States District Court, Northern District of Georgia, Atlanta Division.

First, let's define *"What is a Consent Decree"*?

A consent decree[51] is a settlement that is contained in a court order. The court orders injunctive relief against the defendant and agrees to maintain jurisdiction over the case to ensure that the settlement is followed. Injunctive relief is a remedy imposed by a court in which a party is instructed to do or not do something. Failure to obey the order may lead the court to find the party in contempt and to impose other penalties. Plaintiffs in lawsuits generally prefer consent decrees because they have the power of the court behind the agreements; defendants who wish to avoid publicity also tend to prefer such agreements because they limit the exposure of damaging details. Critics of consent decrees argue that federal district courts assert too much power over the defendant. They also contend that federal courts have imposed conditions on state and local governments in civil rights cases that usurp[52] the power of the state.

---

50     United States District Court, Northern District of Georgia, Atlanta Division, Civil Action File No.1:95-CV-2550-TWT

51     http://legal-dictionary.thefreedictionary.com/consent+decree

52     Take the authority of another by force, without the rule of law.

In the case against the City of Atlanta:

Objective:

To resolve issues alleged by the citizen Plaintiffs, Government Plaintiffs in their complaints for the City of Atlanta to comply with all conditions of its NPDES permits for its CSO Control Facilities, to use sound engineering practices to design and construct any improvements to individual CSO Control Facilities required by the Consent Decree; to use sound management, operational, maintenance practices to implement all the requirements of the Consent Decree; and to achieve expeditious implementation of the provisions of the Consent Decree for the purpose of (1) achieving full compliance with the NPDES permits for the CSO Control Facilities, the Georgia Water Quality Control Act relating to all the City of Atlanta CSO Control Facilities; and (2) eliminate all Unpermitted discharges from the Combined Sewer System.

Rules and Regulations for Water Quality Control-

The Water Quality Standards provide that the following criteria be deemed necessary and applicable to all waters of the State and United States and the discharge must be controlled to prevent the following conditions downstream of the discharge(s):

> All waters shall be free from materials associated with municipal or domestic sewage, industrial waste or any other waste which will settle to form sludge deposits.

> All waters shall be free from oil, scum, floating debris associated with municipal or domestic sewage, industrial waste or other discharges in amounts sufficient to be unsightly.

> All waters shall be free from toxic, corrosive, acidic, or caustic substances discharged from municipalities, industries or other sources, such as non-point sources.

All waters shall be free from materials related to municipal, industrial or other discharges which produce turbidity, color, odor, or other objectionable conditions which interfere with legitimate water issues.

The City shall develop and submit for the EPA/EPD and the Citizen Plaintiff's review and approval, a program for the evaluation of each of the City's CSO Control Facilities.

Remedial Actions for the Combined Sewer Overflow Facilities-

The City shall fully meet and comply with all conditions of the Consent Decree, Clean Water Act, the Georgia Water Quality Control Act, and the NPDES permits for all its CSO Control Facilities to include:

An Evaluation Program to evaluate each Combined Sewer (CSO) Control Facility to determine the overall effectiveness of the existing CSO controls in achieving compliance with the Clean Water Act, the Georgia Water Quality Control Act, NPDES permit for CSO Facilities.

Evaluate the pollutant removal efficiencies of the facilities.

Determine dry and wet weather flows from the Combined Sewer System to the Wastewater Treatment Facilities.

Evaluate the response of the Combined Sewer System to rainfall events. The program will also include whole effluent toxicity testing.

The City agrees to perform the following tasks or their EPA/EPD approved equivalent for each CSO Control Facility in the Evaluation Program.

Describe the Pollutant Parameters to be sampled and analyzed for each CSO Control Facility, which at a minimum shall include ammonia, biochemical oxygen demand, fecal coliform bacteria, total residual chlorine, pH, phosphorous, temperature, total suspended solid oil and grease, total and dissolved zinc, total and dissolved nickel, total and dissolved copper, and any other pollutant parameters in the monitoring requirements of the NPDES permit.

Describe sampling locations for each CSO Control Facility.

Describe sampling procedures, equipment, and analytical methods to be used for each CSO Control Facility consistent with 40 CFR Part 136 (OSHA).

Monitor continuously all flow to the associated Wastewater Treatment Facilities and describe the flow monitoring equipment and its location.

Monitor all flow discharge, including bypass flow, from the CSO Control Facility to the receiving stream.

Describe the techniques and methods for measuring flow of the receiving stream at the in-stream sampling location.

Monitor all rainfall continuously.

Measure a minimum of ten overflows, representative of the intensity and duration of a range of storm events, for each CSO Control Facility during the twelve (12) month evaluation period.

Sample storm water runoff from streets or parking lots, runoff from yards or parks, runoff from two other land used typical of the sewershed, all prior to entering the collection system.

Collect and Analyze twenty four (24) hour flow proportional samples of the dry weather flows to the Wastewater Treatment Facility during two weekends.

Conduct effluent toxicity testing for at least one overflow at each CSO Control Facility.

Describe all hydrological, hydraulic and water quality models which will be used for data analysis.

Describe chain of custody procedures for all collected samples.

Provide a map of the watersheds above the in-stream sampling location(s) of data/information storage for all information collected.

A deficiencies report identifying system deficiencies (design, structural, process, operational, and maintenance) and performance-limiting factors which may limit the overall effectiveness of the existing CSO controls in achieving compliance with the Clean Water Act, the Georgia Water Quality Control Act and the NPDES permits for the CSO Control Facilities.

The City within thirty (30) days of receiving the EPA/EPD's comments shall modify the CSO Control Facility Evaluation Program accordingly, and submit the modified program to the EPA/EPD for final approval.

*Notes:* The objective in short and the consensus by the EPA/EPA and citizens of this Decree was Sewer Separation.

The following parts are excerpts from the Amended Consent Decree that were the issues not addressed or included in the EPA/EPD review of the Consent Decrees on the City of Atlanta as filed in the Federal Court, United States District Court, Northern District of Georgia, Atlanta Division.

## First Amended Consent Decree[53]

Objective:

To resolve issues alleged by the Plaintiffs that were not resolved in the CSO Consent Decree, to comply with all conditions of the NPDES permits for Wastewater Treatment Facilities, to use sound engineering practices to design and construct any improvements to the individual Wastewater Treatment Facilities, Wastewater Collection and Transmission Systems, to use sound management, operational, and maintenance practices to implement all the requirements of the First Amended Consent Decree; and to achieve expeditious implementation of the provisions for the purpose of (1) achieving full compliance with the NPDES permits for the Wastewater Treatment Facilities, Georgia Water Quality Control Act and the Clean Water Atlanta relating to all the City of Atlanta Wastewater Treatment Facilities and Wastewater Collection and Transmission Systems; (2) eliminating all Unpermitted Discharges; and (3) eliminating all Sanitary Sewage Overflows.

Remedial Actions for the Combined Sewer Overflow Facilities

*Maintenance Management System:* City of Atlanta shall install and implement a Maintenance and Management System (MMS) for the R.M Clayton, Utoy Creek, and South River Water Reclamation Centers (WRCs) as approved by EPA/EPD to ensure that preventative and corrective maintenance is conducted and that equipment integral to proper

---

53    United States District Court, Northern District of Georgia, Atlanta Division, Civil Action File No.1:98-CV-1956-TWT

operational and maintenance, treatment is in compliance with the NPDES permits.

Maintenance Training Program

City of Atlanta shall develop a Maintenance Training Program, including an implementation schedule for each Water Reclamation Center to consist of training of maintenance supervisors, system for tracking and training activities.

Operations Program

City of Atlanta shall revise the current operations program for the R.M Clayton, South River, and Utoy Creek Water Reclamation Centers (WRCs) Operations Program to achieve compliance with the NPDES permit to include the following elements for each WRC:

Operations manuals for equipment, descriptions of the operational controls, maximum flow that each process unit may treat before WRC effluent quality is expected to exceed NPDES permit limits, an organizational chart, procedures for the year-round disposal of biosolids in the event the City cannot incinerate all or a portion of the biosolids generated at the WRCs.

> Develop Standard Operating Procedures to include an emergency response plan for abnormal conditions such as power outages and weather related events to achieve compliance with NPDES permits, ensure safety of all personnel, and ensure proper communications of the current operational state of the WRC's.

Laboratory Information Management System

The City shall implement upgrades to the current Laboratory Information System (LIMS) that service R.M Clayton, South River, Utoy Creek Water Reclamation Centers to include the following elements:

Procedures that ensure samples are collected; procedures that ensure sample monitoring instruments and laboratory equipment, quality assurance, and quality control practices that conform to the Handbook for Analytical Quality Control in Water and Wastewater Laboratories and Standard Methods for the Examination of Water and Wastewater, procedures to assure participation in EPA/EPD Discharge Monitory Report Quality Assurance Studies.

Composite Correction Program

The City shall provide for the EPA/EPD with a certification bearing the seal of a Professional Engineer, licensed in the state of Georgia and competent in the design of wastewater treatment facilities, that the planning and design of the Capital Improvement Program fulfills the goals and purposes of the EPA Composite Correction Program as described by the EPA.

Document Retention

The City shall retain records for 3 years at the laboratory for analysis performed, sample data, quality control data, standard curves, calibration and maintenance records of laboratory instruments, process control monitoring records, facility operations, maintenance records, data information used to complete the permit applications. The City shall retain records 5 years discharge monitoring reports and monthly operating reports to include monitoring data regarding the disposal or beneficial re-use of biosolids.

Capital Improvements

The City shall have a Waste Water Treatment Facility Capital Improvement program for the R.M. Clayton, Utoy Creek, Intrenchment Creek, and South River Water Reclamation Centers. The Capital Improvements include construction of new equipment as well as upgrade and rehabilitation of existing equipment. The CIP will upgrade, rehabilitate, or replace sanitary sewers as needed to address sewer overflow. Also under the

First Amended Consent Decree (FACD) were included Operations & Maintenance Contingency Plan, Pre-Treatment Program, and Reporting.

Reporting

The City shall report orally to the EPD within 24 hours all the Unpermitted Discharges to waters of the United States or the States from the City's waste water treatment facilities. The City shall submit to the EPA/EPD within 5 days of the discharge a written report the extent condition, cause of the discharge.

Remedial Actions for Waste Water Collection & Transmission Systems

Consists of a Comprehensive program including the development and implementation of programs to insure proper management, operation, and maintenance (MOM) for the Evaluation and Rehabilitation of the City of Atlanta's Collection & Transmission System and addresses remedial actions for Interjurisdictional Agreements with Satellite Systems and reporting requirements. The MOM program will develop and implement an Emergency Response Plan, a short term Operational Plan, Pump Station Evaluations, a Grease Management Program, a Sewer Mapping Program (GIS), a Maintenance Management System, Training & Safety programs, a Capacity Certification Program, and a Long Term Operations Plan.

The Evaluation & Rehabilitation of the Collection & Transmission System consists of a multi-phased program which will result in a thorough analysis and evaluation of the City's Wastewater Collection & Transmission Systems. The identification of deficiencies, and the correction of deficiencies through three phases:

Phase 1- develop several plans including Flow and Rainfall Monitoring, Hydraulic Modeling, System – Wide Prioritization, and Sewer System Evaluation.

Phase 2- a Macro System Evaluation that assess the major components of the City's Wastewater Collection & Transmission Systems and prioritize sewer basins into six sewer basin groups for individual evaluation.

Phase 3- implementation of the corrective actions identified during the evaluations to include sewer rehabilitation, sewer replacement, and construction of relief sewers. The infiltration & inflow (I/I) evaluations (peak flow) and sewer system evaluation survey should be compatible with EPA's handbook Oct. 1991 or Water Environment Federation's Manual 1994.

The City shall develop and implement a written

1) Collection System Contingency & Emergency Response Plan (CSCERP)

## Timeline

1880- Atlanta's Sewer System

1920- Atlanta's Combined Sewers

1948- Federal Water Pollution Control Act

1970- EPA established. OSHA standards established.

1972- EPA amends Water Pollution Control Act.

1981- EPA amends the Clean Water Act.

1988- City of Atlanta funds CSO Management Study.

1989- EPD issues order mandating Atlanta's CSO's be controlled or eliminated. Georgia legislature passes law requiring Atlanta's water discharges meet water-quality standards.

1990- EPD approves city's CSO facilities plan for screening and disinfection.

1991- EPA handbook guidelines for sewer systems, infrastructure analysis & rehabilitation.

1993- Piedmont Park Master Plan established.

1994- Upper Chattahoochee Riverkeeper established.

1995- Upper Chattahoochee Riverkeeper, downstream property owners file lawsuit against the City of Atlanta.

Mayor Bill Campbell hires consulting firm to address infrastructure problems, CSO issues.

1996- Atlanta hosts Olympic Games.

1997- EPA audits City of Atlanta CSO plants and sewers system.

1998- City of Atlanta enters into CSO Consent Decree. City of Atlanta starts implementing Consent Decree work.

1999- City of Atlanta establishes Greenway Acquisition Plan. Georgia legislature passes Bill 399 for Greenspace Program. Grant Park Master Plan established. US Infrastructure Inc., studies Atlanta's buried assets.

2001- EPA/EPD authorizes City of Atlanta sewer separation plans. EPA/EPD approves Greenway Acquisition Plan- Supplemental Environmental Project (SEC).

2002- Mayor Shirley Franklin's first term starts. The formulation of Department of Watershed Management.

Mayor Franklin's Administrative Order, June 2002. Tunnel pre-designs.

2003- Chastain Park Master Plan established.

2004- Rob Hunter replaces Jack Ravan as DWM Commissioner. Consent Decree construction work of Sewer Separation (storm/sewer) starts.

2006- Mayor Shirley Franklin elected to second term.

2007- Consent Decree (mandated) sewer separation of Greensferry, Stockade, and McDaniel basins completed. Greenway Acquisition Project completed.

2009- Mayor Shirley Franklin's second term ends in December.

2010- City of Atlanta reports Consent Decree work ahead of schedule. First Amended Consent Decree (FACD).

2012- Scheduled anticipated early completion of second Consent Decree (FACD) by the U.S. District Court.

2014- Deadline for second Consent Decree (FACD).

# Part 1 - The Problem

The City of Atlanta's sewer system dates back to the late 1880s. Most drinking-water lines in the United States were installed around 1895. Atlanta's population started growing and the city expanded, tying in a combined sewer system in the 1920s and the system has pretty much stayed the same since. As the city grew,[54] and as sinks, bathtubs, and flush toilets became popular, storm-water sewers became conduits for carrying household wastewater as well. Combined sewers were a common method of sewage disposal in major cities. Once the city outgrew this system, the design and construction of sewers were modernized, resulting in great improvements. Since these systems were not designed to remove waste, and this combined waste was outfalling into the city's streams, they eventually became a threat to public health. As the region grows, the demand on existing infrastructure increases, and the city is faced with greater challenges in managing wastewater and storm water, and in protecting its rivers and streams.

The major problem was that the city was polluting the Chattahoochee River, which is not only where Atlantans draw 100 percent of their drinking water but is also critical for drinking water and recreational purposes in communities all the way to the Gulf of Mexico.

The problem was compounded by the age of the system (deterioration), population growth as a result of the addition of new communities and businesses, inadequate capacity of the system to carry flow or to store

---

54    http://www.cleanwateratlanta.org/overview/History.htm

flow when heavy rains occurred, industrial contamination with grease and chemicals, clogged pipes (dirt, grit, roots, trash, fats, oils, grease, concrete, sanitary products), and broken and missing pipes connecting the system causing inflow and infiltration and with no maintenance, creating a falling system in disrepair.

In dry weather, the collection system conveys all wastewater flow to a water-reclamation center (WRC) for treatment. During wet weather, the combined flows (wastewater and storm water) exceed the collection system's capacity, resulting in combined sewer overflows (CSOs) at the city's six CSO facilities. The CSOs occurred about sixty-plus times per year at the CSO facilities in the west, and about twenty-plus times per year at the CSO facilities in the east. During heavy rainfalls, the peak rate of the CSOs exceeds the screening capacity of the facilities, at which point flow bypasses the screening and causing overflows.

The EPD had determined that Atlanta's sewer system needs to be separated from its storm sewer and gave the city a mandate to complete the separation. The EPD determined that sewer manholes were surging and spilling into nearby creeks, thus allowing sewage to make its way to outfalls and into the Chattahoochee River and posing a health risk. Another mandate has been set for the repair and maintenance of all the basins in Atlanta by area. The city developed the SSES/Rehab Construction Management Team, made up of people from the Department of Watershed Management, the Program Management Team, and Contract Architects and Engineers. This study was done to take a snapshot of one basin (Nancy Creek) in Atlanta to see what condition the concrete pipe was in, what repairs were performed, and the cost over a one-month period.

Industrial plants and commercial businesses were illegally dumping and piping waste directly into pipes that go into the Chattahoochee River.

# 1. Assessing Atlanta's Problem

Assessing the multiple problems that plagued the city's infrastructure was not an easy task. The city had determined that these were the main culprits that needed to be addressed immediately:

- Surging manholes causing sewer spills all over town in the roadways, outfalls, parks, schools, homeowners' backyards, basements, communities all over town, and downtown after heavy rains.

- Damaged roads and sidewalks from leaks beneath the surface

- A complete retooling of bad house connections either damaged, broken, or missing going to the main line.

- Overloaded pipes in the Buckhead corridor, where growth has overtaken infrastructure.

- Fats, oils, and grease from restaurants and businesses were clogging pipes because they didn't have or maintain grease traps.

- Refurbishing the city's drinking-water system and quality of the drinking water.

- Replacing sewer mains, upgrading facility plants, and replacing aging water meters throughout the system.

- Debris blockage in pipes causing backups, overflows, and even collapse.

- Separating the Combined Sewer System from the Storm Sewer System, this completely overwhelms the whole system during heavy rains.

- Finding cost savings to include-

- Taking back the water services that were contracted out to a private company that was mismanaged.

- Taking back utility-locating services once handled by contractors, saving $900,000 annually.

- Taking back bio-solids handling from contractors at four water-reclamation centers saving twelve million annually.

- Taking back CSO sampling from contractors, saving $300,000 annually.

- Handling solids at a savings of $2.6 million.

In addition, bringing up the rear was:

- Flow entering the system through infiltration, and flow exiting the system through exfiltration

- Industrial waste (illegal dumping) into the creeks around the city making its way into the Chattahoochee River.

- Deteriorating structures all over town not allowing for proper collection.

- Flow isolation

- Creating a system wide identification of structures consistent with EPA guidelines.

- Upgrade treatment plants to filter trash from the wastewater and spray it with bacteria-killing chlorine before releasing it into streams.

- Conduct a Sanitary Sewer Evaluation Survey (SSES) to determine condition of piping and structures to identify areas collapsed, broken, deteriorating, clogged and missing pipe, root intrusion, grease intrusion, silt debris, concrete, mud, sticks, lumber, tires, rags, and sanitary products blocking pipes flow.

- Storage capacity and tunnels for storm water and sewage.

- Upgrade treatment facilities.

In a Watershed Management Case Study[55] (part of the Atlanta Case Study Project) written by Harvey K. Newman and Tim N. Todd, they write: hidden infrastructure, such as water and sewer pipes, is crucial to the operations of a city, and its construction and maintenance is essential for a city's growth and prosperity. Since these assets are not easily observed, it is easy for city officials to ignore their maintenance in favor of projects that are more readily noticed by their constituents. Aging and poorly maintained pipes are often passed down from one administration to the next. When the problem can no longer be ignored, the repair costs are usually higher. Atlanta, like many other cities, found itself in this situation in 1998, but it did not develop a comprehensive plan to address the problem until 2002.

The case study continues by assessing a statement of the problem: as Atlanta has grown, its wastewater-collection system has grown to include more than 2,200 miles of piping. The combined sewers represent approximately 15 percent of the city's system. Most of these combined sewers are underneath the center of the downtown area and they are some of Atlanta's oldest and largest pipes. Both the water system and the wastewater system function regionally, collecting and treating wastewater from six wholesale customers—Clayton, DeKalb, and Fulton counties, and the cities of College Park, East Point, and Hapeville. Atlanta residents generate 55 percent of the waste treated at the city's wastewater treatment facilities each day. The wastewater systems serve more than 1.2 million people daily and have a collective maximum treatment capacity of 220 million gallons per day.

In 1995, after numerous unanswered attempts to resolve the issues without resorting to litigation, Atlanta was sued by the Upper Chattahoochee Riverkeeper for its failure to comply with the Federal Clean Water Act in connection with the discharge of pollutants from the city's combined

---

55    http://fiscalresearch.gsu.edu/atlanta_case_study/Watershed%20Mgmt%20Case%Study.
pdf

sewer facilities into the Chattahoochee River watershed. Under Mayor Campbell, the city spent more than four million dollars defending itself against this lawsuit despite widely documented accounts of the problem, both in Atlanta and downstream. In 1997, a Consent Decree was issued by US District Court Judge Thomas Thrash after delays by the city in meeting expected water-quality standards, heightening a sense of distrust between the parties involved.

The EPA subsequently filed a second lawsuit after a five-month investigation into the condition of Atlanta's sanitary sewer overflow problems. This suit led to a second consent decree (the First Amended Consent Decree) against the city. This second summary judgment against the city obligated Atlanta to fix its sanitary sewer-overflow problems and, combined with the initial Consent Decree, mandated what amounted to a complete overhaul of the city's water and sewer systems.

Furthermore, the case study states the struggle of Atlanta's CSOs to keep up with heavy rains. City studies at the time stated that the CSOs overflowed about eighty times a year.

The problems facing the city were not just limited to pollution issues. Decades of neglecting to maintain and properly run the water and sewer systems often left city workers and officials with little to no knowledge about the existence or status of water and sewer pipes. The unsanitary conditions created by the city's decaying watershed infrastructure and the city's lack of administrative control over its watershed led to two lawsuits against Atlanta for the city's failure to comply with the federal Clean Water Act and the Georgia Water Quality Act.

These lawsuits by the Upper Chattahoochee Riverkeeper and federal authorities ultimately led the city to agree to a Consent Decree relating to improvements to the city's CSO facilities to meet water-quality standards by 2007. The court actions also led to a second Consent Decree (referred to as the First Amended Consent Decree) in 1999 requiring the city to eliminate all sanitary sewer overflows in its separated system by 2014.

*Notes:*

The CSO plan that the Advisory Panel would create addressed these issues and gave a more overall and detailed report to Mayor Shirley Franklin.

# Part 2 - The Fix

Atlanta's wastewater collection and treatment systems[56] began taking form during the late 1800s. After much debate, the city started constructing a combined sewer system around 1882. The combined sewers represented approximately 15 percent of the city's system. The combined sewers were located near the center of the downtown area and near some of Atlanta's oldest and largest pipes. By 1917, Atlanta had built three wastewater treatment facilities, had started constructing sanitary sewers, and had installed another 323 miles of combined sewers. The combined sewer system (see definitions) had seven overflow points. Interceptors[57] were constructed to divert dry weather wastewater flows to the treatment facilities.

Atlanta's wastewater collection system has grown to include more than 2,200 miles of piping. The pipe network includes reinforced and unreinforced concrete, cast iron, clay, brick, and reinforced and unreinforced concrete arches. Pipe diameters range from six up to more than 120 inches. About 15 percent of the collection system remains combined; the remaining 85 percent consists of separate sanitary sewer piping.

Atlanta residents generate 55 percent of the wastewater treated at the city's wastewater treatment facilities each day. The wastewater systems serve more than 1.6 million people daily and have a collective maximum treatment capacity of 220 million gallons a day.

---

56    City of Atlanta, Dept. of Watershed Management (Asset Management Case Study), Atlanta's Wastewater System.

57    Downhill location where all other sewers flow.

Wastewater is treated at four wastewater treatment facilities: Intrenchment Creek, R.M. Clayton, South River, and Utoy Creek.

Six combined sewer overflow (CSO)[58] facilities in Atlanta provide preliminary treatment, and chlorination and dechlorination for wet-weather flows from the combined sewer system—Clear Creek, Confederate Avenue, Boulevard, McDaniel, Greensferry, North Avenue, Tanyard Creek.

## Atlanta's CSO Plan

The Combined Sewer Overflow (CSO) Remediation Program was designed to bring the city's combined sewers into complete compliance with the Clean Water Act. The plan was authorized by the US Environmental Protection Division (EPD).

After entering into the first consent decree in 1998, Atlanta implemented a public education and involvement program for citizens and developed a CSO Remediation Plan.

In July 2001, after a three-year process of study and citizen input, the federal EPA and state EPD approved the city's plan to eliminate water-quality violations from combined sewer overflows (CSOs). The city's plan involved a combination of tunnels and separation of selected sewer areas. The city submitted a refined plan to the EPA and EPD that would increase the water-quality benefits of proposed portions of the plan and reduce the lengths of the proposed CSO tunnels.

At a cost of $809 million, the plan required the city to complete the following tasks by 2007:

1) Separation of the Greensferry and McDaniel sewer basins and a part of the Stockade sub-basin known as the Custer CSO basin. This part of the CSO remediation plan will increase

---

58    The discharge of wastewater and storm water from a combined sewer system

Atlanta's separate sanitary sewer network to 90 percent (1,980 miles) from 85 percent of the entire system. The plan will also eliminate two CSO facilities.

The city proposed to construct a deep-rock tunnel storage and treatment system that will capture and store combined flows for the northwestern and northeastern quadrants of the combined sewer network. It would also build additional storage for the combined facilities located in the southeastern quadrant of the combined sewer network. All flows from the combined sewer overflow system will be treated before discharge to the Chattahoochee and South rivers.

2) Reduction of the number of permitted, wet-weather overflows from the combined sewer system from 360 per year to an average of four per year at each of the remaining four CSO facilities. The overflows will be screened, disinfected, and dechlorinated before discharge to receiving streams and will meet water-quality standards.

## Clean Water Atlanta Initiative

Atlanta's main water source is the Chattahoochee River. The river flows southwesterly across Georgia toward the western boarder of Alabama, and continues southward to the Gulf of Mexico. During the 1890s, Atlanta moved its water-supply source from the South River to the Chattahoochee River[59] and constructed a twenty-million-gallons-per-day (mgd) piping station and the current dual pool reservoir for $809,000. The system went into operation September 29, 1893. The system functions regionally, serving all Atlanta residents and Fulton County residents south of the

---

59    City of Atlanta, Dept. of Watershed Management (Asset Management Case Study), Municipality Facts.

Chattahoochee River. On average, more than one million customers are served each day by the system.

Atlanta is located in one of the largest and fastest-growing metropolitan areas in the southeastern United States.[60] The asset-management case study stated that three million new residents have migrated to the area since 1970. The city has approximately 430,000 residents with an estimated daytime population of more than one million each weekday. A recent population forecast indicates that parts of the region could grow by as much as 209 percent by 2030. Demands for water services will increase accordingly.

Based off the Clough (Clean Water Advisory) panel, Clean Water Atlanta[61] (CWA) is a comprehensive, coordinated, long-term initiative to improve the region's water quality. Clean Water Atlanta encompasses water and wastewater infrastructure and improvements to the treatment system that are mandated by Consent Decree, as well as watershed improvement projects that extend beyond the requirements. Clean Water Atlanta represents an unprecedented investment in metro Atlanta's water quality.

For the City of Atlanta, Clean Water Atlanta was created to provide a comprehensive and long-term plan to ensure clean water in our great city for the next generation. The goal of Clean Water Atlanta is to create the cleanest urban streams and rivers in the country within a decade for sewer and water, providing clean drinking water, clean storm water, wastewater treatment, grease management, and combined sewer overflow.

In creating Clean Water Atlanta, a five-point plan was developed with specific objectives, including:

---

60    City of Atlanta, Dept. of Watershed Management (Asset Management Case Study), Municipality Facts.

61    http://www.cleanwateratlanta.org/overview/History.ht.

1) To ensure professional management of the Consent Decree program, and complete the city's Consent Decree obligation two years ahead of schedule.

2) To reduce flooding and pollution caused by storm water. To create a storm-water utility focusing on green-space solutions to flooding and pollution created by polluted runoff.

3) Monitor the water quality of Atlanta's streams and rivers to ensure programs are effective. Hire a full-time director of Clean Water Atlanta, and a full-time finance director charged with effectively managing the $3 billion dollars in construction projects.

4) Eliminate sanitary sewer spills. To achieve 90 percent sewer separation within four years, keeping the city well ahead of schedule on its stated goal of full separation within twenty-five years. To conduct Operation Clean Sewer, which is an aggressive preventative maintenance program aimed at eliminating the one thousand annual spills from the city's sanitary sewer system.

5) Implement a Combined Sewer Overflow (CSO) plan to achieve the highest water quality at the lowest cost within the shortest time frame, and implement a CSO solution that achieves 65 percent better water quality than full separation, and 17 percent better water quality than the current authorized plan. Build a CSO tunnel storage and treatment plant that is the most cost-effective alternative ($834 million).

Drinking water is treated at three water-treatment facilities: the Chattahoochee Water Treatment Plant, the Hemphill Water Treatment Plant, and the North Area Water Treatment Plant, which is jointly owned with Fulton County. The drinking-water facilities meet the average daily demand of 120 million gallons per day (mgd) and have a collective capacity of 246.4 million gallons per day.

Water is distributed throughout the service area through a 2,700-mile distribution system. The system also includes more than twenty-one thousand fire hydrants and 155,000 meters.

In 2002, Mayor Shirley Franklin vowed, "Atlanta will have the cleanest possible water for all its citizens." Clean Water Atlanta was designed to tackle both short-term and long-term issues,[62] and it is the most aggressive and thorough program yet initiated by the city to ensure clean drinking water for all residents, and clean streams and clean wastewater flows for Atlanta and its downstream neighbors. The program weaves together a combination of monitoring and enforcement of water-quality issues and lays out the long-anticipated plan to improve Atlanta's aging sewer system.

## A Long-Term Watershed Monitoring Program[63]

To ensure that the Clean Water Atlanta improvements are achieving the necessary water-quality improvements, the city has conducted a Long-Term Watershed Monitoring Program (WMP) to monitor the water quality before, during, and after the improvements are implemented. Currently several streams within the city are listed in the state's 305 (b) report[64] as partially supporting or not supporting their designated use. City staff and consultants, working with the US Geological Survey (USGS) and Southeast Waters (SEW), have begun setting up monitoring stations to gather biological and water-quality data on metro streams, including Nancy, Peachtree, Proctor, Sandy, Camp, and Utoy creeks, as well as the South River. Forty sites will be monitored initially, with twenty permanent sites over the next

---

62    http://www.atlantaga.gov/media/csoplan_101602.aspx

63    http://www.cleanwateratlanta.org/monitoring/Improv/LTWM.htm

64    www.ga.epd.org/Documents/305b/html

decade. Real-time data is available on the USGS website, www.usgs.gov/ realtime.html.

Oversight will be provided by the Upper Chattahoochee Riverkeeper.

The Watershed Monitoring Program will help the city to document biological and habitat conditions in the city's major watersheds, as well as identify sources of pollution and other impairments. Information gathered from the WMP will also help the city to document stream improvements and any pollutant that can be attributed to program implementation, consolidate water quality program sampling requirements to provide data to support the city's storm-water permitting program, and identify new programs to address streams requiring further action by providing the data necessary to make decisions about the direction of the city's future programs. Finally, the WMP will provide water-quality data available to citizens, adopt-a-stream programs, and other environmental organizations and stakeholders throughout the region in support of public education and watershed stewardship programs.

## Financing the Clean Water Atlanta Initiative

The PowerPoint presentation[65] put together by Rob Hunter of the Department of Watershed Management (City of Atlanta), and Eric Rothstein (CH$_2$M Hill) lays out, by way of graphs and charts, the specific problems the city faced with utilities services and financing the program under the approval of the Atlanta City Council.

*Note:* Not all slides are presented.

---

65     http://www.pueblo.us/DocumentView.aspx?DID=480

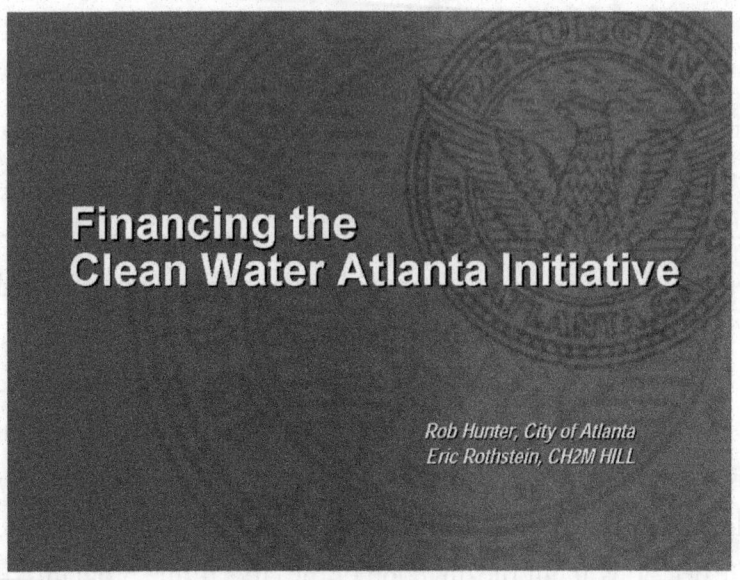

**Financing the Clean Water Atlanta Initiative**

*Rob Hunter, City of Atlanta*
*Eric Rothstein, CH2M HILL*

With the development of the Clean Water Atlanta infrastructure rehabilitation initiative, Mayor Franklin's outline will focus on five key components on which the following slides are based.

## Presentation Outline

- Clean Water Atlanta financial challenges
- 5-year system-wide rate increase plan
- Alternative financing strategies
- Strategic financial planning & financial capability assessment
- Projections of Atlanta's performance
- Summary & conclusions

2

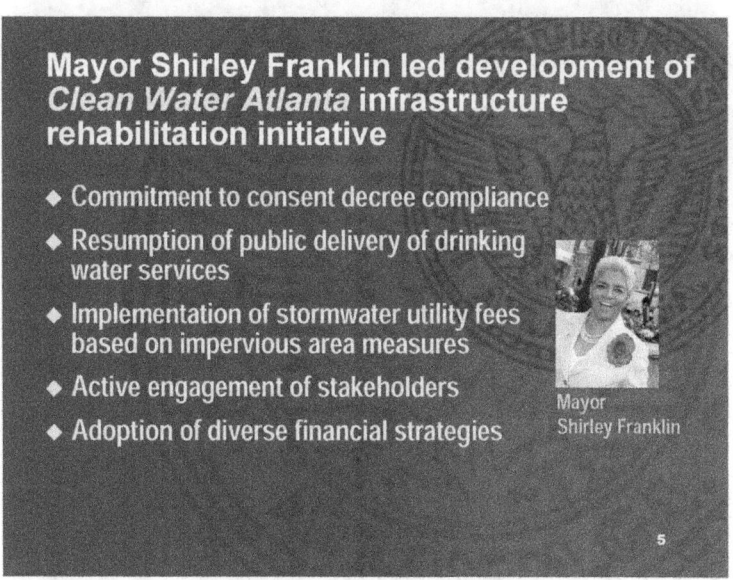

With the following cost projections, rate increases, bonds, tax-exempt commercial paper, and help from the state and federal governments, the

City of Atlanta can reach its objectives of completing the Consent Decrees placed upon it.

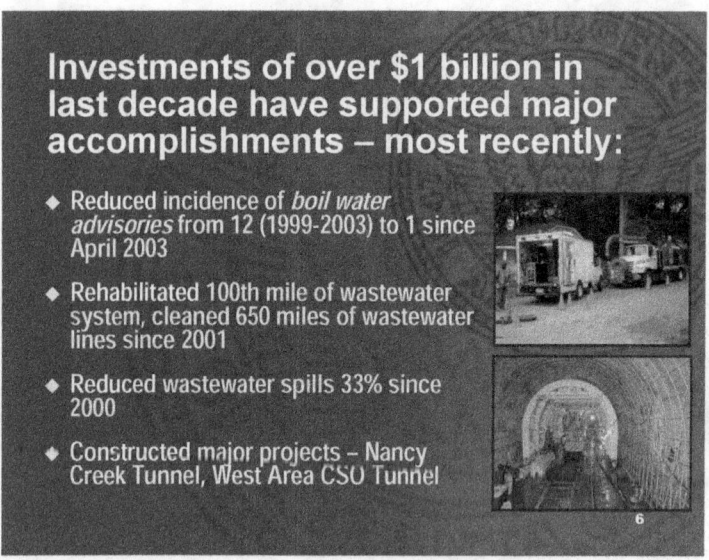

Consent Decree compliance is projected to cost $2.6 billion.

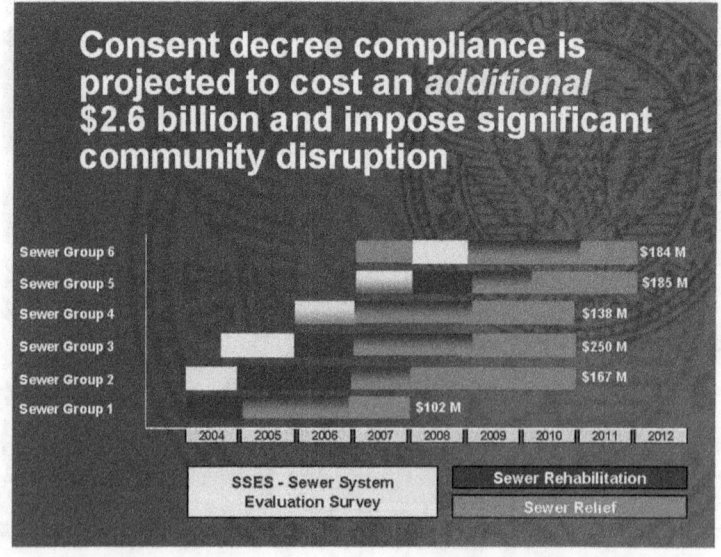

Clean Water Atlanta will require a doubling of the city's water and wastewater fund asset base.

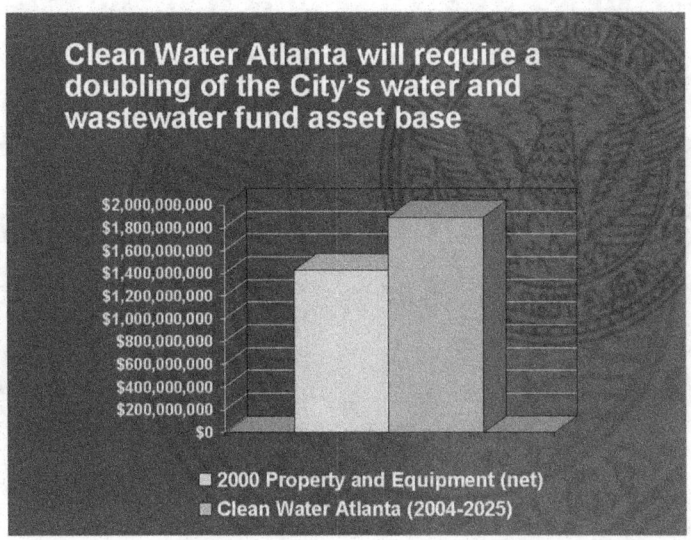

Mayor Franklin's Administration faced a major increase in debt service in 2004.

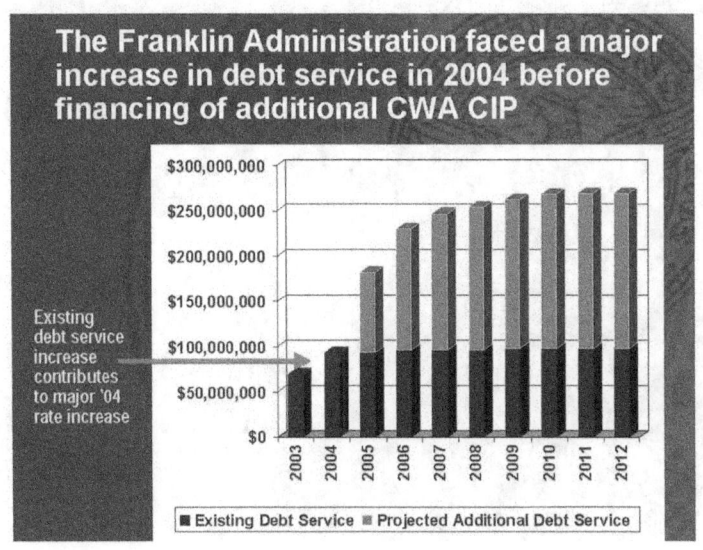

The City will encumber an unprecedented rate increase:

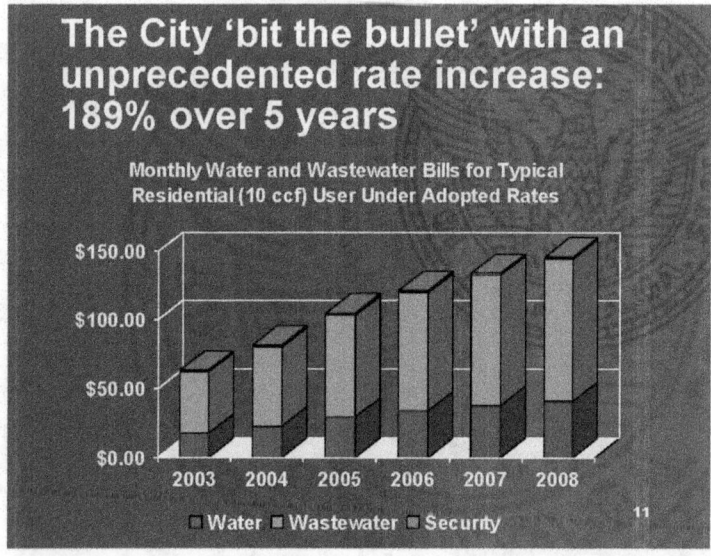

The Rate adoption facilitated issuance of millions in revenue bonds to finance Clean Water Atlanta:

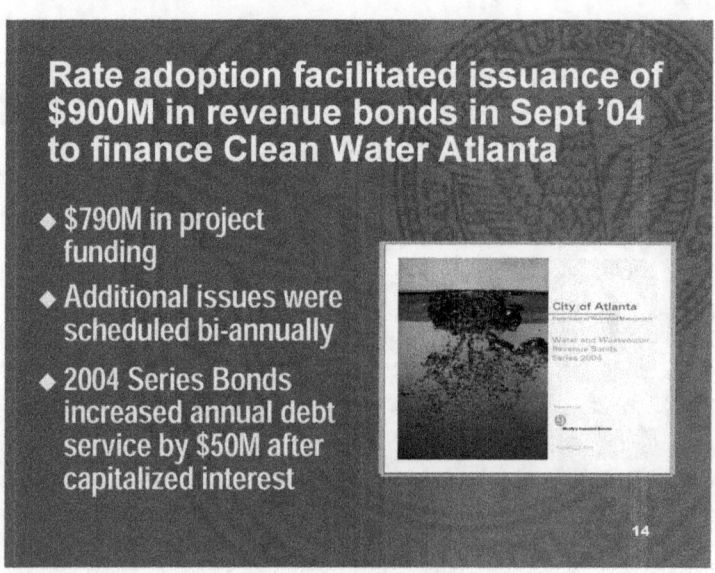

The financing strategies:

**The City has implemented a diverse set of financing strategies**

◆ Municipal Option Sales Tax

◆ Expansion of State Revolving Fund borrowing

◆ Continued pursuit of federal grant funding

◆ Tax-Exempt Commercial Paper Program

◆ Current revenue financing

   → Of recurring capital expenditures

*Balanced use of these measures provides for regional distribution of cost responsibilities while enhancing the financial strength of the City's water and wastewater enterprise funds.*

16

**Atlanta used MOST proceeds to reduce adopted 2005 rate increase 'dollar-for-dollar'**

◆ 1% Municipal Option Sales Tax (MOST) approved in July 20, 2004 referendum

◆ $95M / year for payment of water & sewer capital & operations

◆ In place for 4 years; 2 renewals possible with voter approval

Council tweaks sewer rate breaks

17

**Atlanta's approach to securing financing for the CWA initiative offers important lessons for other communities**

- There is no substitute for clear, honest, and open public communications
  - → Infrastructure as essential for economic vitality
  - → Civic responsibility to honor environmental stewardship responsibilities
- Significant cost impacts suggest use of a broad portfolio of financing mechanisms – rates, taxes, alternative debt instruments
- Political commitment and strong leadership are essential

27

The city received support from the state of Georgia and the federal government:

**The City has also pursued and received support from the State of Georgia and Federal Government**

- Received $4.3 million in grants, primarily for major tunnel projects
- Increased Georgia Environmental Facilities Authority (GEFA) lending from $20M to $50M / year
- Seeking federal line item appropriations

18

The Tax Exempt Commercial Paper plan:

**Tax-Exempt Commercial Paper provides procurement flexibility, reduces arbitrage risk, and saves on borrowing**

◆ Program attributes:
  → $1.2 billion line of credit
  → Short-term interest rates below long-term rates

◆ Program benefits
  → Reduce overall costs of CWA financing
  → Enable cash-flow project financing
  → Facilitate timing of long-term bond issues

**Tax-Exempt Commercial Paper involves active program and debt management**

◆ Monthly projections of project commitments and construction spend-down patterns

◆ Monitoring of short-term vs. long-term interest rates

◆ Timing of conversion of TECP to revenue bonds

Graph numbers:

Vertical bar starting from 0 to $50,000– $200,000 (projected monthly commitments & disbursements)

Horizontal bar starting Dec'05 ending Jan'08

Strategic tools to help evaluate financing strategies:

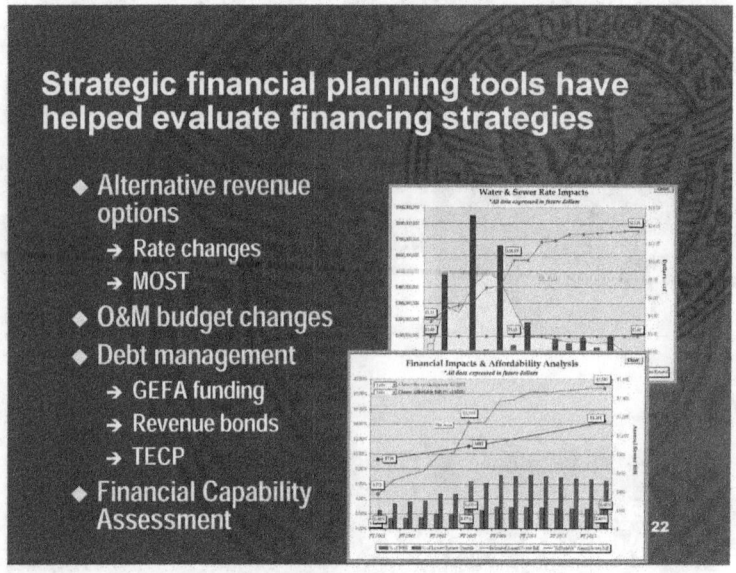

Atlanta's Financial Capability Assessment:

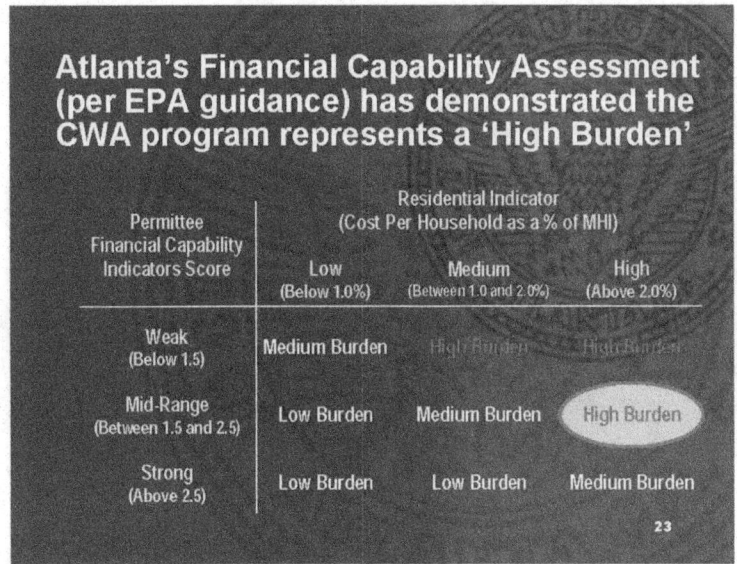

Future rate increases will project a strong financial performance:

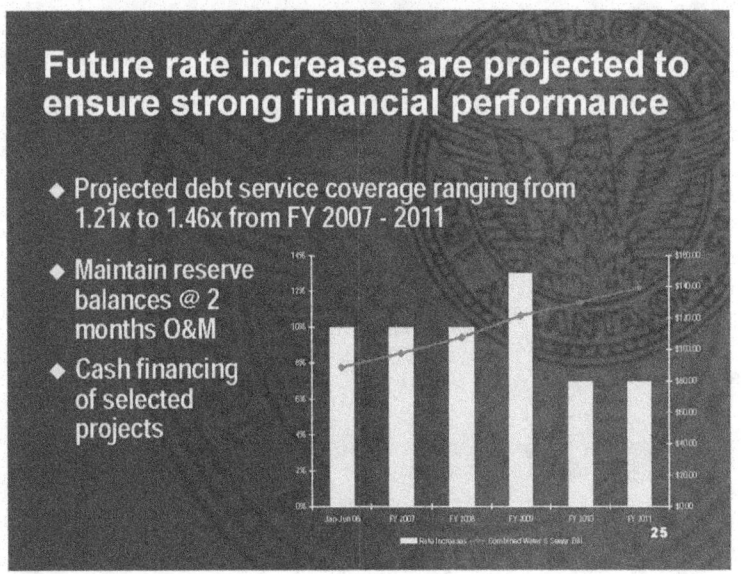

## Schedule of Adopted Water and Wastewater Rates:

| Schedule of Adopted Water and Wastewater Rates With 2005 & FY 2006-2007 Adjustments for MOST Revenues | | | | | |
|---|---|---|---|---|---|
| | 2003 | 2004 | 2005 | 2006 | 2007 |
| Inside-City Water & Sewer | Rate Per CCF | Rate Per CCF | Rate Per CCF | Rate Per CCF | Rate Per CCF |
| Minimum Charge | | $20.49 | $20.70 | $6.00 | $6.60 |
| 0 - 3 ccf | $6.20 | In Min. Ch. | In Min. Ch. | $6.63 | $6.19 |
| 4 - 6 ccf | $6.20 | $7.82 | $7.90 | $8.69 | $9.56 |
| 7 ccf and Above | $6.20 | $8.99 | $9.08 | $9.99 | $10.99 |
| | | % Change | % Change | % Change | % Change |
| 0 - 3 ccf | | 10.16% | 1.02% | 10.58% (for 3 ccf use) | 9.96% (for 3 ccf use) |
| 4 - 6 ccf | | 26.13% | 1.02% | 10.00% | 10.13% |
| 7 ccf and Above | | 45.00% | 1.00% | 10.02% | 10.01% |

29

# Part 3 - The Shirley Franklin Factor

Shirley Franklin ran a strong campaign against her opponents. Because of her knowledge of city government and her experience[66]—having worked under two previous mayors, for Andrew Young as chief administrative officer and as city manager, and afterward for Maynard Jackson as commissioner of cultural affairs, and later as chief administrative officer—the two previous mayors had urged her to run for the position. She had also served five years helping to organize Atlanta's 1996 Olympic Games.

In November 2001, after all the votes were tallied in a tight race, Franklin was the people's choice by winning 50 percent of the votes, and she was elected the fifth-eighth mayor of Atlanta, becoming the first female mayor in the city's history, the first African American woman of a major city in the south, and the fourth African American to hold the title. Her inaguration party was called the People's Inaguration. She wanted people to feel that she was approachable.

## How did she win?

Prior to being elected, she spent two years campaigning. She approached her campaign like the shrewd businessperson she is. Her campaign style was well-organized, to the point, and no-nonsense.[67] Mayor Franklin would say in a *USA Today* article, "All of my female predecessors were well-educated, articulate, experienced African Americans who had been elected to other

---

66    The Biography of Mayor Shirley Franklin

67    www.progressive.org

offices before. I looked at why they didn't win the mayoral race and discovered that they all ran for very short periods of time, like four months. They couldn't remove psychological barriers or raise enough funds in that short time. People have to see you, shake your hand and get to know you. So I spent two years running for office, getting out there and metting the people." Another commitment she took in running her campaign was that of full disclosure. She listed all of her donors on her campaign website and publicly released her last four federal tax returns. Her fund-raising was stunning in its success, beating the funds raised by her nearest political rival, a seasoned politician, by nearly two to one. When the election was over, it was revealed that she had raised and spent three million dollars in her campaign, the most ever spent by an Atlanta mayoral candidate. The Franklin campaign was one based upon embracing various groups rather than focusing only on blacks and women. She stated, "I knew I would appeal to the majority of people who are not anxious at all to return to the time when people didn't get along."

As she campaigned, her indomitable personality would come to light. She became known for her openess and availability to her constituents. A popular bit of campaigning would be the "sleepovers"—late-night get-togethers at supporters' homes, where up to thirty people would hang out and chat with the candidate in a relaxed, casual setting. Though there were dissenters, mainly among well-off business leaders, when election day finally arrived, Shirley Franklin pulled just over fifty percent of the vote—not a landslide by any means, but enough to create history.

*Notes:* Soon after her November election, Mayor Franklin would request consulting firm Bain & Company to prepare an audit report (pro bono) of the city's business model. This was the first of several pro bono engagements by consultants including BCG and numerous Atlanta-based professionals.

In January 2002, Shirley Franklin was sworn into office.

Her first year would turn out to be a very busy year for the newly elected mayor, and she would have to hit the ground running.

After being briefed on the state of the city's finances, she determined that the business model of city government was not working, and the city's projected expenses for the 2002 year would exceed projected revenues.

She had inherited an eighty-two-million-dollar budget deficit[68] (20 percent of the entire city budget) and thirty-two million dollars more than projected. She found out that the sewer-system infrastructure was failing, the pipes were broken and corroded throughout the city, manholes were surging and sending sewage into the streets, backyards, businesses, basements, outfalls, creeks, and into the river. The CSOs (Combined Sewer Overflows) were inadequate to handle the volume of flow that treats wastewater, water meters were broken and not reading correctly, thousands of gallons of water were leaving the system through infiltration[69] and exfiltration,[70] and to top it all off, millions of dollars of fees were not being collected. United Water had signed a contract to provide privatized water operations on January 1, 1999, and things weren't going well on that front either. The city did not know that a high percentage of treated water was not billed and probably leaked out between the treatment plant and the meters.[71]

Also, what newly elected Mayor Franklin may not have known is that in the previous year, 2001, the deputy commissioner of public works was heavily involved with the Georgia EPD in response to the civil action filed against the city to correct the city's CSO (Combined Sewer Overflow) facilities at Clear Creek, Greensferry, and North Avenue and to put in place

---

68    http://www.time.com/magazine/article/0,9171,1050265,00.html

69    Leakage or enrty of sewage or water flow into pipes from the surrounding area (ground) through cracks, roots, and breaks in pipes.

70    Leakage or exit of sewage from the wastewater system into the surrounding area (ground) through leaks in pipes, joints, etc.

71    http://gppf.org/article.asp?RT=20&p=pub/Water/atlanta_water.htm

an emergency response plan to control sewer spills in the city that were going into the Chattahoochee River.

Her strategy would be bold, ambitious, and aggressive. She knew she would
have to make tough choices, that she would be criticized for them, and that
she would be held accountable to the city, the voters, and the federal and
state agencies. Mayor Franklin would have to restore the trust of the seat
as Atlanta's new mayor. She stated, "We want an honest, transparent administration." The significance of her approach would be twofold: to stop
the financial hemorrhaging the violations were causing, and to return the
money to upgrade the system that had resulted in fines being paid to the
EPA. It had become clear to her and everyone involved in Atlanta's politics
that the city's government was anything but lean and mean. Ethics emerged
as issue number one in the wake of the exit of the prior administration.

## Why Ethics?

Ethics evolved out of allegations overshadowing the previous administration's policies. Mayor Franklin knew vigorous action was needed to restore
openness, honesty, and transparency back to the city and to regain the trust
of its citizenry.[72]

Shirley Franklin, used the campaign slogan "I'll make you proud" as a way
of promising to clean up city government. While campaigning, Franklin
announced her intention to appoint groups of experts and citizens to work
on problems facing city government. The first of these was the Ethics Task
Force, which Franklin appointed in December 2001, less than a month
after winning election. Franklin established a goal for the Ethics Task Force
to instill a "culture of ethics" within city government and charged the
members to recommend changes in the city's ordinances. The Ethics Task

---

72      http://fiscalresearch.gsu.edu/atlanta_case_study/Ethics%20Case%20Study.pdf

Force, chaired by a former acting US attorney, Dorothy Kirkley, held its meetings away from city hall and without the mayor's presence.

## Herself as City Ambassador

By holding press conferences and media briefings monthly, her aim was to use the media to inform residents and businesses of the progress the city would be making to transform itself into "Best in Class." The city's compliance with the federal Consent Decree and with federal and state mandates made upgrading sewer and water infrastructure a top priority from day one.

Mayor Franklin and her staff became global ambassadors for city relationships, a government working fourteen-hour days, meeting endlessly with department heads, and openly discussing the city's problems and challenges with a refreshing candor. The mayor, an accidental politican, as she called herself, would throughout her term(s) attend numerous functions to speak at groundbreaking ceremonies, plaque dedications, ribbon cuttings, receptions, certificate awards, black-tie functions, fund-raisers, church engagements, the supermarket, prayer breakfasts, schools, universities, graduations, the Atlanta Rotary Club, and on radio and TV talk shows, including C-SPAN, and NPR. She would also make guest appearances and attend unveilings, luncheons, annual conventions, public briefings, televised debates, park tours, motorboat tours along the river, press conferences, lecture series, training sessions, committee meetings, city council presentations, photo ops, panel discussions, public service announcements, videos on demand, the mayor's website, and make short movie vignettes, all in an effort to show her dedication to achieving her vision for the City of Atlanta.

Mayor Shirley Franklin made repairing Atlanta's sewer system a primary focus, and she used the project to heighten the public's knowledge about and commitment to rebuilding the city's water infrastructure.

## Year by Year

How Mayor Franklin started out 2002:

January

- From her inaugural speech[73] she stated: "It is an honor because I stand on the shoulders of all the great leaders that this city has produced. Today, we embark on a new, bold and daring journey for this state. This is the first time that a woman has been sworn in to serve as Mayor of Atlanta. But Atlanta has always been a bold innovative city. A city ahead of its time; a city not afraid to set the pace because we have been blessed with progressive leadership. And here we stand on the brink of a whole new era of leadership, and I am proud to lead the way. I envision an Atlanta as a city with clean air and clean water, beautiful parks and a vibrant arts community, where people care about their neighbors, where elderly have adequate housing, where children are assured a quality education and healthcare, where corporations provide fair stable employment, and lead the way in facilitating economic development, where religious leaders teach us how to live harmoniously with dignity and respect for all ethnic groups, races, and cultures. In the Franklin Administration, we will embrace this newly diversity in true Atlanta fashion. I want to see Atlanta as a shining example of livable, loveable and workable community."

- Mayor Franklin announced in her first press conference that the city faced a budget gap[74] in an analysis of the city's business model from consultants Bain & Company (a global strategy consulting company) of between seventy and ninety million

---

73    City of Atlanta Online http://www.atlantaga.gov/media/inspeech_010102.aspx

74    http://www.atlantaga.gov/media/budpress_010902.aspx

dollars and listed five principal reasons for the shortfall. 1) The city spent more money than it had budgeted. 2) The city incurred unforeseen expenses as a result of the September 11 terror attacks. 3) Revenues for the current year would fall below original expectations because of the economy. 4) The 2002 budget included full-year funding for commitments made during 2001. 5) The city had to lower the revenue expectations previously reported on prior years.

She vowed to start by cutting her own salary[75] by forty thousand dollars and cutting her own office budget by 30 percent. She directed department heads, external agency directors, and city employees to return all city-owned cars, light trucks, and SUVs assigned to individuals for nonfield administrative purposes. She also recalled all keys and cell phones in her commitment to cut waste and abuse from the city government.

• Mayor Franklin met with the business community by holding a breakfast, stating her commitment to the business of good governance strengthened by cooperative public and private partnerships at the local, city, state, and regional levels. She stated at the breakfast that people love Atlanta, that she was one of them, and that her focus was simple—to make Atlanta a safer, cleaner place; to have a more responsive and effective government, a better city for families, seniors, and children, and above all an open and honest city hall. She continued that the first order of business was a comprehensive review of the proposed 2002 budget, and from the findings of the independent audit performed by Ernst & Young, the firm reported that Atlanta had some of the worst bookkeeping practices the

---

75    http://www.atlantaga.gov/media/paycut_012802.aspx

auditors had ever seen, and many of the finance department employees were clearly unqualified for their jobs.[76] She told the group she had a plan of action and the first step was to find out what the problem was, and step two was to call together call the expertise within city government and without to look at the options. Her goal in the first one hundred days was to raise the standards for ethics in city government, fill key positions to lead the departments, develop a balanced budget and plan for subsequent years, implement a pothole strike team, launch a complaint and information hotline to report complaints about city services, and lastly, step up collections of funds due the city. She closed by telling them, "I'm in it to win."

- Mayor Franklin announced new appointments, starting with Lynnette Young, hired as chief operating officer (COO) tasked with quickly identifying underlying organizational issues contributing to poor service delivery, finding out how departments were operating inefficiently and who/where the deficiencies existed. Lastly, she was charged with collecting meaningful performance data on departments and employees and set goals. She was tasked with fixing the city's broken services. Young challenged departments[77] to find innovative new ways to do more with less.

As COO, she would provide policy direction and executive management over the departments of aviation, police, fire, corrections, parks, recreation and cultural affairs, planning and community development, public works, watershed management, procurement, information technology, human resources,

---

76   http://www.ask.com/wiki/Shirley_Franklin

77   http://aysps.gsu.edu/New_Efficiencies_Case_Study.pdf

office of enterprise asset management, office of contract compliance, and the emergency management services.

- Mayor Franklin would appoint Rick J. Anderson as interim chief financial officer of the city to lead the budgeting process, while search firm Ray and Brendston would seek a permanent officer.

*Note:*

Rick J. Anderson had served two previous mayors as CFO, and his experience made him the choice of the committee for a permanent appointment. He was tasked to "clean up the financial mess" at city hall.

- Mayor Franklin would appoint Greg Pridgeon as chief of staff to act both in a managerial and advisory capacity (confidante) in support of the mayor.

  He would provide advice and guidance to the mayor on legislative and political issues and serve as liaison to the public and private sectors in fostering partnerships and the development of revenue-generating initiatives, as well as providing supervision and policy direction over external affairs and international relationships, constituent services, communications, human services, as well as festivals/special events.

- Mayor Franklin would appoint Sandra Walker as communications director, Greg Gionnelli as senior policy advisor, and David Edwards as senior management officer.

*Note:*

In her first infrastructure move, Mayor Franklin announced the creation of the "Pothole Posse."

Mayor Franklin would declare war on potholes and utility cuts in the streets left by contractors that were not completely covered throughout the city. Through the Pothole Posse, street operations would dedicate seven street-repair crews to track and fill all potholes with the goal of completely eliminating the existing inventory of potholes for the next six months, identifying and repairing over five thousand potholes.

February

- Mayor Franklin hosted "Mayor's Night" in an initiative she started where citizens and employees can come and have a ten- to fifteen-minute sit-down and discuss their concerns or complaints about the city. These meeting had to be screened and scheduled by appointments.

March

- Consulting firm Bain & Company issued result audits of the city's business model to Mayor Franklin from its initial investigations in November that had very specific, detailed, and troubling information about the research it collected.

- Mayor Franklin signed an administrative order implementing a strict ethics policy for all employees who were under the supervision and direction of the mayor to provide mandatory ethics training for the leadership of city government, including all officials and employees at the level of bureau head, office director, and above. The order would include a hotline to report any ethics violations.

April

- Mayor Franklin, the acting commissioner of the department of public works, and civic environmental groups kick off a Storm Drain Stenciling Program that involved printing a message next to each street drain throughout the city to remind the public that litter and other pollutants that enter storm drains often end up in local streams and creeks. The goal was to raise citizen awareness of the connection between storm-drain run-off and water-resource protection.

May

- Atlanta Regional Commission sponsors Leadership, Innovation, Networking, and Knowledge (LINK), where representatives (business leaders, government agencies, and charitable groups) from thirteen counties in the Atlanta metro area visit other regions and meet with their counterparts annually to learn how things are done in other major cities and to try and foster a regional attitude of cooperation and understanding among people who can make a difference. The trip would take them to Chicago because Chicago provided an excellent laboratory of how the mayor of a major city can strike working relationships with other governments in the region was well as with the business community. Atlanta would learn three key lessons from Chicago: 1) how a strong mayor can impact a city, 2) the importance of fixing the city's infrastructure, and 3) the incredibly tight-knit group of people in the business community who work with city hall. Tony Landers, ARC's director of community services, who organized the annual trips, stated, "Chicago is a much older city than Atlanta, and we have the opportunity to figure out how they learned to face challenges over a long period of time."

June

- Mayor Franklin initiates a review[78] of United Water's contract,[79] signed in 1999, which was a twenty-year, $20.8

---

78    City of Atlanta Online http://www.atlanta.ga.gov/media/waterrelease_061002.aspx
79    http://www.serconline.org/waterPrivatization/fact.html

million deal (some said sweetheart deal) making it the nation's largest public-private partnership contract to run the
city's drinking-water system. United Water was contracted
to cover water utility operations, management, maintenance
(Operations & Maintenance), and various services and supplies such as laboratory work, meter reading, and supplying
chemicals. A letter was sent by the mayor to United Water following preliminary findings from an independent committee
she appointed about the dissatisfaction with its performance
under the contract, and plans the city had of determining
whether the contract should be continued.

- Mayor Franklin announced a panel of experts to advise the city
on Atlanta's one-billion-dollar combined sewer overflow plan.
Consisting of nine members and chaired by Wayne Clough, the
president of Georgia Institute of Technology (Georgia Tech),
the panel was called the mayor's "Clean Water Advisory Panel."

July

- The City launched Wireless Broadband Initiative to expand IT
services for Atlanta residents, businesses, and visitors.

- The Clean Streams Task Force submitted a report prepared
for the mayor's Clean Water Advisory Panel titled "Full Sewer
Separation with Stormwater Greenways, An Opportunity to
Revitalize the Face of Atlanta." In the report, the task force
applauded the city's efforts on complying with the Consent
Decrees and the effort the city was undertaking in sewer separation, treatment, and tunnels. *But,* conversely, it was critical
that the city recognize that failure to integrate storm-water
greenways into the sewer separation scenario would only exac-
erbate the water-pollution problem in Atlanta, and the panel

stated that, operationally, greenways were inextricably tied to water quality. When considered in this manner, the combination of greenways with sewer separation must be viewed as a unified whole, and not simply as additional or optional green space and ponds.

September

- Mayor Franklin met with the chairman and CEO of United Water Inc. He told the mayor that it was critically important for United Water to have an effective monitoring program to be implemented during the ninety-day correction period.

- With the city council's approval, Mayor Shirley Franklin created the Department of Watershed Management (DWM) to consolidate the drinking water, wastewater, and storm-water functions that were previously organized in different departments within the city.

- Mayor Franklin's search firm Russell Reynolds Associates announced Jack Ravan as the nominee to head the newly created Department of Watershed Management in 2002. He was picked because of his knowledge and wealth of experience working in watershed management, as a leader in the treatment of chemical waste, and as the EPA's regional administrator for the southeast region.

The Department of Watershed Management established a series of programmatic responses[80] to issues relating to the Consent Decree to include: 1) Financial Program including the development of a financial and capital improvement plan as well as a revenue requirement analysis; 2) Revenue Program including

---

80    City of Atlanta Online http://www.atlantawatershed.org/images/wm_org_051004.pdf

the procurement and installation of a new billing system called a Customer Information System (CIS), new collection policy and procedures, and a reorganization of the billing and collections staff; 3) Customer Service Program focusing on identifying and meeting customer service expectations, including the development of a new call center and customer service staff; 4) Operating System Performance Program consisting of a systematic review of drinking-water and wastewater performance metrics, policies, and procedures, and an implementation plan for improvements in performance and reliability; 5) Water Loss Program including the sequential evaluation and implementation of projects for meter leaks, service leaks, valve and hydrant testing and repair, leak detection, and water main leak repair; 6) Consent Decree Program consisting of an integrated compliance, engineering, construction, monitoring, and reporting system for the wastewater Consent Decrees.

- Mayor Franklin spoke to the city council about emergency flooding[81] throughout the city. She addressed how the city had put in place a readiness response plan to deal with major flooding going forward and when to declare a state of emergency for public health and safety, to implement a disaster assistance plan to give flood disaster assistance (FDA), and to develop a master list of impacted residences.

October

- Mayor Franklin announced her Clean Water Initiative, which would encompass five points: 1) professional management of the consent decree program; 2) a strategy to reduce flooding and pollution caused by storm water; 3) SSO consent decree

---

81    City of Atlanta Online http:www.atlantaga.gov/media/floods2_092302.aspx

compliance; 4) water quality monitoring to ensure the effectiveness of Clean Water Atlanta programs; 5) and CSO consent decree compliance.

- Mayor Franklin announced the "Save our Sewers" Campaign Rally, which would consist of write-in cards to report "It's Broken, Let's fix it." Also that month, the watershed commissioner and the mayor had to explain to residents in the Cascade Road, Fairburn Road, and Harbin Road areas why water service was so poor and lacking on the southwestern side of town.

- Mayor Franklin, along with the city council, approved the launch of the Department of Watershed Management.

- Mayor Franklin and Commissioner Jack Raven, responding to complaints about water-main and valve breaks in two different areas of southwestern Atlanta, announced that the city was working diligently with United Water to restore complete water service in the vicinities of Cascade Road, Fairburn Road, and at Harbin Road. Mayor Franklin promised the city would conduct a thorough review of the situation.

November

- Mayor Franklin presented the 2003 proposed budget to the Atlanta City Council which totaled $426.3 million.

December

- Mayor Franklin met with the governor, Sonny Purdue, to discuss ways for funding infrastructure from the state level.

- Mayor Franklin ended the year as at the beginning, with Franklin's administration setting the course for the next year. In January, the mayor would host the annual mayor's State of the City Business and address the city council, laying out the challenges ahead for the city to emerge as a "Best in Class" city.

Professing herself "The Sewer Mayor," she took the task to hand and called all hands on deck to clean up the creeks, provide improved drinking water to the citizens, and protect the environment in the city, which in turn would benefit downstream communities. "Invest or die" for the life of the city's infrastructure. "I have called our underground infrastructure 'our buried treasure,'" she stated. No city can maintain healthy economic development or a high quality of life if its sewer and water infrastructure cannot provide both sewer capacity and clean, safe drinking water. "The sewers were crumbling to the point that the federal government decided to make an example of Atlanta."

Year in Review

Mayor Shirley Franklin would spend the year focused on restoring the confidence in the city government as being ethical, open and honest; pointing to leadership starting with the mayor, and defining her administration by selecting the best talented staff to help her accomplish her goals; opening up her office to city employees and private citizens to let them know she was listening; selling her program to the public, business, and the state to make them aware of the seriousness of the city's aging infrastructure; developing the Clean Water Advisory Panel (comprising national and local experts) and deciding on the best possible recommendations to address the Consent Decrees, finding a commissioner for the newly created Department of Watershed Management and top-quality people to be in charge of running the program, defending the choice made by the Clean Water Advisory Panel as the best option, stopping the bleeding from fines imposed by the Consent Decree order, and being prepared to pitch the cost for embarking on the 2007 and 2014 deadlines. If Atlanta could meet its current three-billion-dollar price tag, that would mean seven thousand dollars per man, woman, and child in Atlanta or nearly twenty thousand per household or twenty-six thousand per paying customer. Each customer would pay $2,100 a year over twelve years to pay off the cost. Atlanta has no way to pay for it beyond taking out loans and raising rates.

In an October 16 press conference regarding the Clean Water Atlanta campaign, as a part of Operation Clean Sewer, Mayor Franklin would state, "I am today directing Commissioner Jack Ravan to put in place an action plan under which the City will meet its Sanitary Sewer Consent Decree obligations two years earlier than required under the federal court order." "For too long the City has responded to urgent problems by seeking delays. Clean Water Atlanta is an action program, and I will not allow us to seek delays in meeting our clean water obligations. So I am committing to the EPA and EPD, to the city council, to the many environmental groups joining me today, and to every resident of Atlanta, that we will finish all our Consent Decree requirements by 2012," Franklin said at the time.

Dr. Wayne Clough, the chairperson of Clean Water Advisory Panel, acknowledged that the panel's recommendation may be less than ideal because the city faced a 2007 deadline as part of the federal Consent Decree. With more time, perhaps the city could have drafted a better solution that would have included more green space, water features, the burying of utility lines, as well as improving streets and sidewalks.

Jack Ravan, the commissioner, stated he would revisit the financial estimates and look for ways to lower costs. Further, while Atlanta may not have the time to create an ideal CSO plan by the 2007 deadline (the first deadline), it was critical that all people involved be open to revisiting the panel's recommendation so we could implement the best possible plan for now and the long term.

Mayor Franklin and her staff would take several trips to Washington, DC, in hopes of raising as much as one billion dollars from the federal government over the next decade.

Awards

- ❖ 2002 Certificate of Achievement for Excellence in Financial Reporting to the City of Atlanta- Government Finance Officers Asociation of the United States and Cananda.

- ❖ 2002 GFOA's Distinguished Budget Presentation Award. (Government Finance Officers Association)

In order to be awarded these certificates (valid for one year), the government had to publish an easily readable and efficiently organized Comprehensive Annual Financial Report. This report satisfied both Generally Accepted Accounting Principles (GAAP) and applicable legal requirements. The City of Atlanta, Georgia (Comprehensive Annual Financial Report for the fiscal year), December 31, 2002, presented by Mayor Shirley Franklin and Chief Financial Officer Richard J. Anderson.[82]

The report states that "we believe that our current CAFR continues to meet the Certificate of Achievement Program's requirements and we are submitting it to the GFOA to determine its eligibility for another certificate. In order to qualify for the Distinguished Budget Presentation Award, the government's budget document was judged to be proficient in several categories, including as a policy document, a financial plan, an operations guide, and a communications device".

The report summarized the independent audit involved in examining, on a test basis, evidence supporting the amounts and disclosures in the financial statements; assessing the accounting principles used and significant estimates made by management; and evaluating the overall presentation

---

82     http://www.atlantagov/client-resources/government/finance/

of the financial statement. The independent auditor rendered an unqualified opinion that Atlanta's financial statements for the fiscal year ended December 31, 2002 are fairly presented in conformity with GAAP. The independent auditor's report is presented as the first component of the financial section of the report.

Critics

Opposition would come from city neighborhood planning groups, citizens, business owners, and even members of the city council, saying they did not trust Atlanta, which would present challenges for the new administration. Several people questioned the financial and pollution analyses given by the panel as flawed, and the recommendation did not weigh the amount of green space that would be a part of the option as an important criterion for Atlanta's future quality of life. Mayor Franklin cut one thousand jobs from the city payroll, setting off a firestorm within the city government. City officials also feared along with rising property tax bills, the rate hikes would make the city unaffordable for many lower- and fixed-income residents. Activists who monitored activities at city hall showed little interest at all. Activists said the building of the two deep storage tunnel system would pollute groundwater and won't solve the problem as well as separating sewers from rainwater. They wanted Atlanta to fully separate sewers from rainwater and build a series of ponds to retain rainwater.

In 2003

January

- Mayor Franklin and the City of Atlanta sought to terminate the city's contract with United Water Inc,[83] citing hundreds of complaints by homeowners about brown water and poor

---

83      http://www.academia.edu/2801672/Meeting_the_mandate_for_clean_water_
an_evaluation_of_privately_managed_US_water_and_wastewater_systems; case study
Atlanta,Georgia,ppg 29-46

service. She noted that the problems included staffing levels, bill collection, and meter installation and repair. Furthermore, United Water was seeking an additional eighty million dollars[84] for services the company claimed were provided outside of the contract.

*Note:*

A performance audit by the Office of the City Internal Auditor found that United Water's collection performance fell short of the 98.5 percent performance standard included in its contract. The poor performance caused the water and sewer accounts receivable to more than double. Additionally, the contractor was not shutting off services or charging late fees on delinquent accounts.

The purpose of a performance audit is to provide information to improve public accountability and facilitate decision making. Performance audits encompass a wide variety of objectives, including those related to assessing the effectiveness of programs and results; economy and efficiency; internal control; compliance with legal or other requirements; and objectives relating to providing prospective analysis, guidance, or summary information.[85]

• The State Department of Transportation stated that United Water failed to repair recurring leaks that had been problematic for years and the company did not make the repairs until the state threatened to hire contractor(s) to fix the problem and backcharge United Water.

---

84    http://gppf.org/article.asp?RT=20&p=pub/Water/atlanta_water.htm

85    Comptroller General of the United States, Government Auditing Standards, Washington, DC: US General Accounting Office, 2003, p.21

- Mayor Shirley Franklin went on a quest to find money to fund the multibillion- dollar rehabilitation of sewers and upgrades for drinking water and water treatment plants . She was persistant in meetings with city council members, Fulton County commissioners, governor Sonny Perdue, the Georgia General Assembly, and Congressman John Lewis to get funding. Mayor Franklin knew that a 1 percent sales tax or special sales tax dedicated to water and sewer infrastructure projects may be her best option. She made it known to everyone that it was a dilemma and the task ahead was a local, state, and national crisis that required a local, state, and national solution. Mayor Franklin also let it be known that unless Atlanta received assistance, water and sewer rates would increase dramatically in 2004.

February

- The city has developed a comprehensive revitalization plan for the Vine City area. The purpose of this was to move swiftly and expeditiously to alleviate the effects of flooding in the Vine City community. The plan would restore this neighborhood by developing a comprehensive plan combining waste and storm-water treatment with community relocation and redevelopment. Residents had to be displaced in September 2002, and community outrage had grown. The plan offered local homeowners the option to go into brand-new homes in the Vine City area and would provide relocating assistance to renters who lived in the affected areas.

- Recommendation made to the city council to move forward with purchasing seventy residential and commercial properties identified as "flood prone" and request the city council to authorize that this new project be funded with existing bond revenue.

April

- City of Atlanta ends its contract with United Water.

May

- Commisioner Graves developed a plan for the Vine City comprehensive revitalization to include: 1) an offer by the city to purchase the impacted homes for up to 120 percent of their preflood, fair-market value; 2) a provision of down-payment assistance of up to fifty thousand dollars to those residents who decided to relocate in the Vine City community; 3) assignment of responsibility to the community development corporation to absorb the cost of the lots to hold down overall costs, as the lots were initially bought with city funds; 4) and assignment of responsibility to build replacement housing to the two Vine City-based community development corporations working in conjunction with the city. The home replacement program, coupled with the city's ongoing efforts to rehabilitate less severely impacted homes, was designed to help stabilize the Vine City community and to encourage homeowners to remain in the area.

- Mayor Franklin sends a letter responding to congresswoman Denise Majette's disapproval of the proposed option for the CSO (Combined Sewer Overflow) plan in October 2002, stating the refined plan was incorporated with the CSO plan and authorized by the Environmental Protection Agency and the Georgia Environmental Protection Division on January 22, 2003 and approved by the court. She closed the letter stating all solutions to the CSO problem were expensive and all must be implemented in the near term.

- Mayor Franklin, Bernie Marcus, Douglass Draft, governor Sonny Perdue, major contributing corporations, and other dignitaries were on hand for the official groundbreaking for the Georgia Aquarium[86] and the new World of Coca-Cola site downtown by Centennial Olympic Park. She said the project would create a major cultural, educational, and entertainment destination for the people of Atlanta and visitors from around the world and was to be the top Atlanta attraction and venue for Atlanta events in the heart of downtown.

June

- Mayor Franklin announces Donna Owens as the deputy commissioner of solid waste services. The mayor said Owens had extensive experience in the field of solid waste and served with the Washington, DC, Department of Public Works.

July

- Launch of the City of Atlanta's new Clean Water Atlanta website. The site is dedicated to detailed coverage of Atlanta's $3.4 billion water and wastewater infrastructure program. The site offers information about the program, and material on: 1) wastewater, storm-water and drinking water projects; 2) Consent Decree programs and compliance schedules; 3) the city's public involvement and information programs; 4) and links to the city and the Department of Watershed Management. The site contains "Hot Topics" and "Hot Projects," information about

---

86    www.georgiaaquarium.org

ongoing work and news of interest to the community and a public-education section with an interactive "Kids Corner."

- Council members, Jack Ravan, commissioner of the Department of Waershed Management, and community leaders celebrated the completion of the Peachtree Trunk–South Fork Relief Sewer Projecton on July 17.

*Note:*

The new sewer line was capable of handling the current flow and anticipated commercial and residential development in the area. It would prevent sanitary sewer overflows into surrounding creeks and prevent backups into neighborhood homes and yards. The infrastructure project involved the replacement of a sewer line that was too small to handle the current flows. The new sewer line crosses South Fork Peachtree Creek three times, two in the vicinity of Cheshire Bridge Road, Lenox Road, the CSX Railroad, and I-85 via tunnels. The project ends at Briarcliff Road and South Fork Peachtree Creek.

August

- Mayor Franklin, and council members unveiled the City of Atlanta's new Trash Trooper Program in conjunction with the Keep Atlanta Beautiful/Great Clean-Up. The program involves a strike force of workers who are assigned around the clock and tasked with providing quick response to community cleanup requests that cannot be served by regular-hour crews, who provide daily scheduled collections of garbage, yard trimmings, and bulk rubbish.

- Completion of Phases I–IV of the Tenth Ward Trunk Sewer Relief and Rehabilitation project. The improvement project involved the replacement of a sewer line too small to handle its current flows.

- City of Atlanta celebrated the completion of the first one hundred miles (528,000 linear feet) of sewer rehabilitation by holding an event at Chastain Park. The city's sewer rehabilitation program is ahead of schedule and under budget.

*Note:*

The city had rehabilitated over one hundred miles of existing sewers, exceeding its goal of twenty-four miles per year by 25 percent. Before Mayor Franklin took office, less than ten miles of sewer lines were being rehabilitated each year. The project had exceeded its goal of averaging more than thirty miles per year. The completion of this stage of the rehab project means that 1/16 of the city's sanitary sewer system fourteen completed sewer sheds, were 100 percent rehabilitated. Jack Raven, the commissioner, stated that "we are justifiably proud of this work. With this milestone, we are on our way to fulfilling Mayor Franklin's pledge of a world-class sewer system."

September

- Mayor Franklin sent a letter to the Georgia EPD and the EPA responding to the July 18, 2003, request by the EPA and the EPD in connection with the First Amended Consent Decree (FACD) made in connection with an investigation of the cause of the city's failure to meet the Indian Creek Trunk Relief Sewer project deadline (February) and monitoring progress toward completion, stating the causes of compliance failure and corrective actions related to the Indian Creek sewer construction project. The City of Atlanta committed to specific corrective actions in order to ensure that future projects would meet

Consent Decree milestones. The Indian Creek project in the Pine Hills area called for replacing 16,700 square feet of sewer using tunnels that were dug at varying depths about thirty feet below ground.

- Office of the mayor and the Program Management Office issued a powerpoint presentation to the mayor showing the results from the Citizen Satisfaction Survey that they conducted. The objectives were to: 1) create a citizen satisfaction index against which the performance of city services could be measured; 2) Drive an "outcomes orientation" into the city management culture; 3) and provide the mayor and the COO with a reliable channel that captures the shifting priorities and attitudes of citizens that can be used to drive service decisions and business strategies.

October

- Mayor Franklin sent a letter to governor Sonny Perdue requesting state assistance. She stated in the letter, unless the city received assistance, water and sewer rates would increase dramatically beginning in January 2004. Residential, commercial, and wholesale customers would receive a 45 percent increase in 2004, another 45 percent increase in 2005, and an 11 percent increase in 2006, 2007, and 2008. She went on to request the following: 1) the state of Georgia and the General Assembly approve grants to the city of fifty million dollars for ten years; 2) the governor to assist her and her lobbying team to urge Georgia's US Congressional delegation and the Bush administration to provide one billion dollars in appropriatations to the city for relief from the mandate; 4) and support proposed legislation by the Georgia Municipal Association (GMA)[87] that would allow a Municipal Option Sales Tax (MOST)[88] at the earliest possible date. Mayor Franklin commented that along with a commitment from the state, a partnership with the federal government was entirely justified. The city must, and would , meet the requirements and deadlines imposed under the Consent Decree negotiated by the federal government.

- Mayor Franklin, Jack Ravan, the watershed commissioner, communitity leaders held a press briefing and photo op inside the Nancy Creek Tunnel to celebrate the one-year anniversary of the Clean Water Initiative.

November

- City council refused to pass legislation to raise water and sewer rates.

December

- Mayor Fraklin met with governor Sonny Perdue and Sam Williams[89] of the Metro Chamber of Commerce to discuss a public-private partnership to develop a plan to tackle the $3.2 billion sewer-improvement project. The result of Williams's involvement resulted in $500 million in state support toward the city's sewer improvements.

- The Georgia Association of Black Elected Officials passed a resolution supporting Mayor Franklin's efforts to find money to repair and expand Atlanta's infrastructure.

---

87    www.gmanet.com

88    https://etax.dor.ga.gov/salestax/atlantamunitax.aspx

89    http://www.metroatlantachamber.com/content/IntPage.aspx?Id=158&SId=2

- Mayor Franklin launched the Blueprint to End Homelessness. The 24/7 Gateway Center is a safe haven for homeless Atlantans that provides showers, toilets, food, storage, and beds. Services include medical/mental health care, job readiness, job training, and specialized services for veterans, women, and children.

- Commisioner Jack Ravan sent a letter to the EPA and the Georgia EPD responding to a notice of delay to the First Amended Consent Decree (FACD), citing the Atlanta City Council's approval of an ordinance increasing water and sewer rates and a Department of Watershed (DWM) ordinance. The letter stated that these ordinances differed significantly from the proposed rate increase and budget requests put forward by the city's administration as a component part of meeting the terms and conditions of the Consent Decrees. Specifically, the approved rate increase ordinance substanially diminished the effective rate increase sought of 45 percent for the fiscal year 2004 by providing an affordable base rate exemption of 600 cubic feet per month for all customers at the current rate. The budget ordinance cut twenty-five million dollars from the DWM operations budget for 2004. The 600 cubic feet rate structure as passed by the council undermined the financial strenght and stability of the water and wastewater fund. Specifically, it resulted in a revenue loss of twenty-five million dollar in the first year, which compounds to over seventy-five million dollars by the fifth year and also results in a declining debt coverage ratio. Consequently, this rate structure does not provide for a solid five-year financial plan or revenue stream that could adequately demonstrate to Wall Street that the city could support any future borrowing.

Year in Review

Mayor Franklin spent the year focused on reviewing United Water's contract,[90] having auditors pick apart flaws and poor performance by the company. She sent the company a letter demanding immediate improvements to the system, as well as payment for what she said was United Water's failure to collect more than eighty million dollars in fees the previous year. Franklin said United Water was part of the problem because it had insufficent staffing. She said the company had fallen behind in collections and meter installations. Franklin's letter to the company, which amounted to a warning, stated that things had to change if United Water were to keep its contract with the city.

Despite United Water's response to the complaints, Franklin's administration said the company would have ninety days to prove that it should keep its twenty-year, forty-million-dollar contract.

Mayor Franklin would end the contract and give the management of water services to the Department of Watershed Management.

She took her campaign for fixing the sewers and water system to the streets. Focused on funding the massive overhaul, she compared the task of raising federal funding from Congress to rolling a snowball uphill in July. "This is a very difficult task and a complex project," she said. "We have very little time to make our case." The city had to meet construction deadlines by 2007. Senators Saxby Chambliss and Johnny Isakson had both expressed strong support for the Atlanta sewer project.

Mayor Franklin and the DWM got in a scramble starting in February to complete the Indian Creek project. Contractors were required to work twenty-four hours a day, seven days a week to complete the project.

---

90    http://www.11alive.com/news/loca/story.aspx?storyid=18094

Mayor Franklin had tried to get federal, state, and county assistance for Atlanta to help pay for some of the sewer infrastructure projects, but she had been turned down at every corner. The absence of outside help would increase the average homeowner's monthly bill by about eighty-seven dollars in January 2004 if the council approved the recommendation.

The Franklin administration proposed a five-year series of rate increases to fund improvements in the wastewater and drinking-water systems. The city council passed a package of significantly lower rates. Mayor Franklin vetoed that legislation and informed the Environmental Protection Agency that the city would be forced to default on its federal Consent Decrees. Bond-rating agencies downgraded the city's water and sewer fund-bond rating. Rate negotiations between the administration and the city council in December resulted in a rate package substantially equivalent to the administration's original proposal.

Kasim Reed, a state senator, was the sponsor for the Municipal Options Sales Tax (MOST) sales tax legislation in the General Assembly. The MOST was authorized by the General Assembly.

The city launched the Clean Water Atlanta website; focused on revitalization throughout the city; the City of Atlanta would celebrate the completion of the first one hundred miles of sewer rehabilitation and sewer relief projects.

Mayor Franklin had successfully found funding for the $3.2 billion infrastructure project with the help of Sam Williams and governor Sonny Perdue.

Mayor Franklin announced the New Century Economic Development Plan, which focused on economic and infrastructure development.

The Georgia Association of Black Elected Officials (GABEO) passed a resolution supporting Mayor Franklin's efforts to find money to repair and expand Atlanta's infrastructure.

The City of Atlanta passed an ordinanace requiring that all new and major renovated, city-financed construction projects (over five thousand  square feet or over two million dollars in cost) would  at a minimum incorporate sustainable design criteria, and the design and project-management teams were  required to meet LEED™ Silver-certified level.

*Note:*

By enacting this ordinance, the City of Atlanta would  incorporate sustainable building design and construction practices into city-financed projects and promote consistent application of sustainable green building practices.

The city approached the Senate HUD/VA subcommittee for a $150 million appropriation earlier in the year to help fund the Nancy Creek Tunnel under a Federal Support Initiatives.[91] The city succeeded in  getting a line item New State Funding commitment of two million dollars from the Senate subcommittee and continued to lobby for additional federal support. The funding strategy:

> Special Local Option Sales Tax (SLOST) $500 million
>
> Low Interest State Loans-      $500 million
>
> Federal Appropriations-      $1 billion
>
> Ratepayers-      $1+ billion

*Notes:*

Evelyn Lowery, Juanita Abernathy, and  Andrew Young, , who envisioned bringing a Center for Human and  Civil Rights to Atlanta,[92] approached Mayor Franklin. The recommendation was a center  to commemorate the contributions of Atlantans and Georgians to the historic struggle for African American freedom and equality. The center would  also serve as a

---

91     http://www.cleanwateratlanta.org/overview/Funding/Federal.htm

92     http://www.atlantadowntown.com/initiatives/center-for-civil-and-human-rights

space for ongoing dialogue , study, and contributions to the resolution of current and future freedom struggles of all people at the local, national, and international level. The idea was originally concieved by Maynard Jackson during his second term as mayor. Jackson saw a continuation of the civil rights movement through a focus on politics. Jackson believed that achieving social change did not rest in violence but in using legal powers and political inroads that civil rights activists in the 1960s were unable to achieve.[93] As mayor, he pushed for affirmative-action programs that ensured black-owned businessess received a proportionate number of municipal contracts, and he worked to alleviate poverty among Atlantans. After Jackson's death on June 24, 2003, Mayor Franklin described his unwavering commitment to using electoral politics to effect social change.

In an article[94] by Cathy Woolard, president of the Atlanta City Council, in the *Atlanta Journal-Constitution*, she writes that the "challenges on sewers must be met." In brief, the Atlanta City Council and Mayor Franklin were not as far apart as it seemed on the next steps to solving our water and sewer problems. The cost of doing nothing or not doing enough would mean even higher bills to constituents in the long run. We have no other choice but to raise rates dramatically, as painful as that may be. The water and sewer crisis is complicated. There are large past bond payments and future bond payments to consider. We are still absorbing the reacquired water department. We are transitioning work that was being done by high-priced consultants into full-time staff positions. Every adjustment to the finances in one column affects the numbers and programs in others. That's why reaching a compromise was so difficult. The council and administration needed to create a framework for making these tough decisions together. We needed to establish a schedule for decision making, establish agreeable

---

93    http://www.civilrights.uga.edu/cities/atlanta/mayor_maynard.htm

94    *Atlanta Journal-Constitution* archives: Cathy Woolard, Dec 13, 2003

norms for communications and recognize that it often takes more than two rounds to agree on a compromise. We needed to think creatively about new ways to fund our infrastructure improvements. Ending, she stated: "My hat goes off to this council and Mayor Franklin for their willingness to venture into areas that most politicians prefer to avoid. Despite the challenges, future Atlantans will be better off for their efforts."

Critics

Critics questioned the real $3.2 billion price tag for infrastructure work.

The watershed department had a reputation of getting numbers wrong for ratepayers all over Atlanta, stated Mary Norwood.

Money was in short supply[95] as the federal government racked up huge deficits to fund US involvement in Iraq and Bush's war on terrorism.

Atlanta's sewer overhaul threatened to triple sewer rates over the next decade if federal and state aid couldn't be found.

Atlanta's Georgia delegation didn't rank strong as it had a few years before. Michael Binford, a political scientist at Georgia State University, stated, "I think the city's chances of raising one billion dollars are slim."

At an original cost of $250 million from the Home Depot co-founder Bernard "Bernie" Marcus,[96] as a present to Atlanta announced in November 2001, Marcus drew up plans to build the Georgia Aquarium in downtown Atlanta. It would be the world's largest aquarium, with more than 8.5 million gallons (32,000 m³) of marine and fresh water and housing more than 120,000 animals of 500 different species.

---

95 *Atlanta Journal-Constitution* archives: D.L. Bennett, June 23, 2003 http://docs. newsbank.com/s/InfoWeb/aggdocs/NewsBank/0FBDAB42A91EDCC9/0D57227E12704560 ?p_multi=AJBK&s_lang=en-US

96 http://en.wikipedia.org/wiki/Georgia_Aquarium

With cost overruns and to get the reinforcing bubble glass on time for the displays, the final cost would come in around $290 million. Some say it was a savvy political move by the businessman to get the site secured for the new georgia aquarium, a development that forced Mayor Franklin to pass and enforce an ordinance on pandhandling, loitering, homelessness, and sleeping in Centennial Park past closing hours to clear out the area. The ordinance also took away free parking on the streets for three to four blocks around the site.

Governor Perdue rejected Mayor Franklin's formal request[97] for millions in state aid to help pay for upgrading Atlanta's sewer system in a letter to her, saying, "It is impossible for me to commit the state to grants of $500 million over ten years to the city."

Eric Johnson, the Georgia Senate President Pro Tempore, said he was frustrated that Mayor Franklin had sent him a letter instead of calling him or visiting his office at the Capitol to ask for his support. Johnson said he would not burden American taxpayers with Atlanta's problems by appealing to federal officials. "We didn't cause Atlanta's problems and we shouldn't have to bail them out, no more than we have the right to ask Atlanta to give us a half billion for our budget problem."

The councilwoman Mary Norwood is the only member who had said on record she would not vote for a rate increase.

At Indian Creek (near Lenox Square Mall), city officials ran into a series of construction delays because of difficult rock and soils and the failures of tunnel-boring machines. The project ran 139 days behind. As a result, the around-the-clock construction caused grief and a bombardment of complaints about the noise and traffic. Critics wonder if Indian Creek, which

---

97     *Atlanta Journal-Constitution* archives: Ernie Suggs, October 28, 2003 http://docs. newsbank.com/InfoWeb/aggdocs/NewsBank/0FE78A6B733BE265/0D57227E12704560 ?p_multi=AJBK&s_lang=en-US

was one of ten short-term projects require under the first Consent Decree, would be a sign of things to come.

Grumblings about all the sewer work led critics to question what the city was doing about affordable housing, schoolteachers, police officers, firefighters, and other civil servants who could not afford to live in the communities where they work in, stating, "we are a community of haves and have-nots, and the folks making the rules continue to pretend their rush to please every interest group with a gripe is not contributing to this great divide."

*Note:*

Because of the court order, if the council voted <u>no</u> on the increases or tried to get an extension of when infrastructure work could be done, a federal trustee would be appointed to come take over the sewer system.

Mayor Franklin had said to the city council, "We have run out of time." The city council would either vote for it or the federal government would step in. And when that happens, "I am going to stand up and say I told you so."

A standoff ensued between Mayor Franklin and city council members who refused to pass legislation that would have significantly raised water and sewer rates to pay for court-ordered repairs.

In 2004

January

- The city imposed the first in a series of rate increases to fund capital and system improvement programs adopted unanimously by the city council. Mayor Franklin stated that "with this vote, we will be able to meet the deadlines of the program and all EPA and EPD consent decrees and court orders."

- In January's billing to customers, they would begin to see the impact of the new rate structures, which would provide sufficient revenue to cover bonds that would fund the overhaul of the city's antiquated infrastructure (Fig 3).

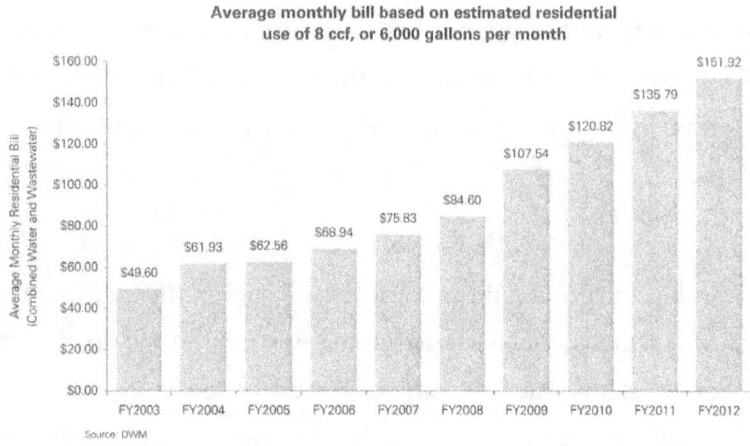

Average monthly bill based on estimated residential use of 8 ccf, or 6,000 gallons per month

- Keep Atlanta Beautiful and seventy students from Atlanta International School (AIS) began a new project to tackle the issues of litter prevention and environmental awareness on a citywide basis.

- The City of Atlanta had to close the Mitchell Street Bridge indefinitely near the Richard B. Russell Federal Building in accordance with a letter received by the State Department of Transportation citing the critical condition of the bridge.

- Mayor Franklin and city officials led Congressman Oberstar on a tour of the Nancy Creek Tunnel.

February

- Mayor Franklin vetoed a budget amendment to decrease appropriations to water and sewer funding, saying, "It is the City's obligation to provide clean water. This obligation is mandated by the Clean Water Act, Safe Drinking Water Act, the 1998 and 1999 Consent Decrees, and the drinking water Consent Decree." She further told the council, "As Chief Executive Officer for Atlanta, I would be irresponsible if I allowed this reduction in the budget."

March

- City Officials and the Ben Hill Community celebrated a ribbon-cutting opening of the newly renovated Melvin Park Drive Baseball Field.

- Mayor Franklin launched "Welcome to Atlanta–NASCAR Weekend"

- City of Atlanta received proposals for the sale and redevelopment of City Hall East (*see future projects chapter*).

April

- City of Atlanta issued Notice-To-Proceed on the West Area CSO Tunnel project.

May

- Mayor Franklin and the Department of Watershed Management celebrated National Drinking Water Week and the one-year anniversary of Reclaiming Atlanta's Water System.

- Mayor Franklin and Community Partners held BeltLine press conference.

- The Campbellton Road Widening Project began.

- Mayor Franklin at a press conference reaffirmed the city's commitment to a dollar-for-dollar rollback of water rate increases should Atlanta voters approve the 1 percent Municipal Option Sales Tax (MOST) on July 20.

June

- The City of Atlanta announced the departure of Jack Ravan and introduced the deputy commissioner, Rob Hunter, as the new commissioner of watershed management.

- A street cave-in forced the shut down of Auburn Avenue between Bell Street and Fort Streets.

- As a part of Mayor Franklin's continuing effort to repair and eliminate potholes, the Department of Public Works (DPW) announced a plan to fill 1,500 potholes in six weeks on arterial and collector streets within a one-mile radius of Peachtree Street and Martin Luther King Jr. Drive.

July

- Mayor Franklin, council members held a press conference to thank the people of Atlanta for approving the Municipal Option Sales Tax (MOST). She stated that "with the passage of the MOST, we share the burden of paying for the rebuilding and upgrading of the City's water and sewer system with the people who visit. We are on the way to achieving the common goals of improving our water and sewer system, minimizing the rate impact to our customers and providing clean safe water for future generations."

- TSA completed installation of three new check baggage security screening systems at Hartsfield-Jackson International Airport, totalling $215 million.

- Mayor Franklin welcomed drug czar John Walters to Atlanta, where she had implemented Weed and Seed program to combat the war on crime in the city.

- Mayor Franklin and council members held a press conference to "Set the Record Straight" stating Atlanta's crumbling water and sewer system must be fixed. She urged Atlantans to stick to the facts: the $3.2 billion Clean Water Atlanta program had been thoroughly reviewed by the independent engineers, the Environmental Protection Agency, the State Environmental Protection Division, federal courts, and nationally renowed experts, including the Clean Water Advisory Panel and the deputy executive director of the American Society of Civil Engineers (ASCE) at the City's request. She went on in the statement to say that a recent ad in the *Atlanta Journal-Constitution* proclaimed that "the potential $1.79 billion identified by the Fulton County Review and Analysis (in conjunction with Parsons Engineering) deserved a closer look." See said that nothing could be further from the truth.

- Pryor Street reopened to traffic after a five-month closure because of construction.

August

- The City of Atlanta began implementing the Community Improvement Project as part of the Quality of Life Bond Program. A $150 million Quality of Life Bond Program leveraged funds to add new sidewalks, beautify major arterials, enhance both pedestrian and vehicular safety by improving traffic signals, trails, and amenities, as well as funding the reconstruction and resurfacing of damaged streets and bridges.

September

- Mayor Franklin announced a task force for a "Walkable Atlanta." The goal was to develop a vision of what Atlanta would look and feel like when walking throughout the city to residents, commuters, and businesses. The focus will be on street design, street safety, sidewalk design, and sidewalk maintenance.

October

- Mayor Franklin and volunteers lent a hand to clean up graffiti in Atlanta during volunteer day. Also in October, the city introduced a groundbreaking ceremony for Cascade Road Streetscapes (phase I) to begin installing new sidewalks, crosswalks, and pedestrian lighting, all in an effort to enhance pedestrian safety and to improve the appearance of the community.

- Mayor Franklin, the Department of Public Works, and council members celebrate the groundbreaking ceremony for the Cascade Road Streetcape (phase I).

- Sidney Marcus Boulevard City Repaving Project from Piedmont Road to Linburg Lane.

- The Municipal Option Sales Tax (MOST) effective October 1, is a 1 percent City of Atlanta sales tax that will be collected for retail sales and use occurring within the incorporated city limits of Atlanta. The purpose of the tax is to assist with funding renovations to the water and sewer system.

November

- Atlanta City Council members, David Scott, the commissioner of the Department of Public Works, and the Kirkwood Community celebrated the groundbreaking ceremony for the Hosea Williams Streetscape (phase I).

- Mayor Franklin received the 2004 Public Officials of the Year honor.

She would comment that "she figured that if she governs well, the voters will be smart enough to know it. People want someone who'll just tell it like it is."

Year in Review

Mayor Franklin spent the year focused on how to present the rate increase to the citizens of Atlanta as well as the city council.

In 2004, Rabbi Lindenblatt, with the help of Mayor Franklin, the Atlanta City Council, the DeKalb County Commission, Georgia Power, and BellSouth, established an eruv[98] in the Morningside and Virginia-Highland neighborhoods[99].

In 2004, Hartsfield-Jackson Airport set a new international air-cargo tonnage record with 349,000 metric tons of international cargo passing through the city.[100] The tonnage increase broke the previous record set in 2003 for international tonnage by more than sixty-six thousand metric tons.

In May, the Atlanta BeltLine was announced as the most comphrehensive economic development effort ever undertaken in the City of Atlanta and among the largest, most wide-ranging urban redevelopment project to be undertaken in the United States.[101] Some government watchdogs are not impressed. They see the Beltlline as a project fraught with overspending. The Fulton County Taxpayers Foundation, in a newsletter dated September 2011 and titled "The Atlanta BeltLine: Is It Feasible?" claimed that BeltLine officials widely overspent on the purchase of a small piece of land in Atlanta's Old Fourth Ward. In an article titled "Did BeltLine pay $3.5 million for less than ¼ acre?" the group states that "several parcels

98    http:www.universitycityeruv.org/WhatIsAnEruv.html

99    http://anshifard.com/Eruv/eruv.html

100   http://www.investatlanta.com/atlCompAdvan/transportationHub.jsp

101   http://www.politifact.com/georgia/statements/ 2011/sep/28/john-sherman/did-beltline-overspend.htm

were purchased by Georgia Power from the City of Atlanta and then sold back to the Atlanta BeltLine Inc., for seemingly huge profits."

By the end of the year, many residents would endure a 45 percent increase in their water bills. Mayor Franklin would write a letter in June to the citizens' advisory board that if the sales tax passes, the rate increase for next year would be reduced from 43 percent to about 13 percent.

Despite MARTA, the city would still need mass transportation.[102] The proposed Atlanta BeltLine, pushed both by Mayor Franklin and Cathy Woolard, the city council president, would create a transportation corridor along underutilized and abandoned railroads to serve Atlanta's in-town neighborhoods. The BeltLine would connect more than forty historic Atlanta neighborhoods and cross more than four thousand acres of industrial land suited for green space and mixed-use development. However, railroad companies, the owners of the right-of-way, were less than eager to see the development started. A study of the idea had been turned over to MARTA.

*Note:*

Mayor Franklin's most ardent political rival, city council president Cathy Woolard, would give an "A" for the performance of the mayor so far.

In 2005

January

- Mayor Franklin reported a revenue surplus of eighteen million dollars.

- Mayor Franklin kicks off Spring Street Water Main Replacement Project

---

102   http://www.georgiatrend.com/April-2004/Madam-Mayor-to-the-rescue

- Mayor Franklin celebrates New Sister City Partnership Between Atlanta and Fukuoka, Japan.

February

- Mayor Franklin hosted a reception program and exhibit in the City Hall atrium to honor African-American Pioneers in Water and Environmental Sciences in commemoration of Black History Month.

- The Department of Watershed Management brought a new Customer Information System Customer Call Center (911) and phone system online and Customer Management. With this new system in place, the agency anticipated a dramatic reduction in the number of estimated bills and resolved customer billing issues. Customer service representatives could now access multiple data systems to provide comprehensive answers to customers' questions.

- Sewer rehab project starts in Buckhead.

March

- Mayor Franklin initiated the Clean Street Teams, a city initiative that employed homeless persons to remove litter from Atlanta's streets and interstate ramps.

- City of Atlanta launched Large Meter Testing Project.

- Mayor Franklin, commissioner Rob Hunter, lead city employees of DWM to volunteer in stream cleanup.

- Atlanta ranked number two among top ten tech cities.

April

- The final phase of roadway construction around the airport for the TSA baggage screening project (Terminal North) nears completion.

- Boring begins for the Flint River Force Main Project.

May

- The City of Atlanta is named a top "mid-sized city" arts destination.

June

- City of Atlanta, Department of Parks Recreation and Cultural Affairs present "Park Design Tennis Court Renovations."

July

- Mayor Franklin and council members celebrate the ground-breaking ceremony for the McDaniel Sewer Separation/Main Replacement Project.

August

- Mayor Franklin announced the Greenway Acquisition to protect communities green space.

- Mayor Franklin toured the Chattahoochee River with the Upper Chattahoochee Riverkeeper and vowed "a cleaner, safer Chattahoochee."

- Construction began on the Peachtree Corridor Sewer Project.

- Fairburn Road mains replaced.

- Mayor Franklin and officials address the military's recommendation to close Fort McPherson and what roles plans the city and the state would play in planning the redevolpment.

- Mayor Franklin and council members celebrate the groundbreaking ceremony for the Stockade Sewer Separation Project.

September

- In association with the American Red Cross, the City of Atlanta set up a relief center at the Adamsville Recreation Center on MLK Jr., Drive in southwest Atlanta to serve as a primary site for emergency responders and as a transition center for evacuees displaced by Hurricane Katrina. The center provided job placement, student services, food stamps, advice on Medicare and Medicaid services, a general store, and Internet services. Volunteer physicians from Grady Hospital and Emory University were on-site for medical assistance.

- The City of Atlanta began Peachtree Corridor Sewer Project, Phase I (24/7 construction).

October

- The City of Atlanta completed Phase I of Peachtree Corridor Sewer Project in record time.

- Mayor Franklin and council members celebrate the groundbreaking ceremony for the Greensferry Sewer Separation/Main Replacement Project.

- Mayor Franklin and council members kicked off the Virginia-Highland Main Replacement/Sewer Rehab Project.

November

- Mayor Franklin launched Wi-Fi services at Hartsfield-Jackson Internation Airport.

- Mayor Franklin announced the Fort McPherson Planning Local Redevelopment Authority (LRA).

- Mayor Franklin and the Department of Watershed Management recommission the Chattahoochee Fountains.

December

- Mayor Franklin hosted an official celebration for the startup of the Nancy Creek Tunnel and pump station, which had been under construction for three years. The tunnels' function was to convey sanitary sewage to the treatment plant (R.M. Clayton Reclamation Center) and eliminate sanitary sewer overflows in the Nancy Creek basin. As part of the festivities, Mayor Franklin pushed a ceremonial button to launch fireworks marking the introduction of flow into the tunnel. This project was the first major consent project completed on budget and on schedule.

- Mayor Franklin is chosen as one of *Esquire* magazine's "Best and Brightest."

*Note:*

This aggressive approach since she was elected in 2002 had earned Major Franklin a second term in 2006 with 90 percent of the votes. She was re-elected uncontested and continued to plow forward to meet consent decree after consent decree even to the astonishment of the federal judge overseeing the Consent Decrees. Mayor Franklin had given the people what they asked for—a restored government accountability to the people, and strengthening the effectiveness and efficiency relating to public services. By the end of Mayor Franklins first term, she was credited with turning an eighty-two-million-dollar deficit into an eighteen-million- surplus.

Year in Review

"If you look in a book on how to get re-elected, its kind of like the not-to-do lists," stated Mayor Franklin.[103] In her first term, she had shown Atlantans a no-nonsense, back-to-basics policy making based largely on broad public-private partnerships. As a long time city administrator, she focused more on selling policies and shoring up the basic systems of government—such as instituting walking beats for Atlanta's police—than on winning political points, though she was quick to say she didn't sit on the political sidelines. "I like politics, now don't get me wrong, but I don't believe in playing politics with government policy," she said. "We ought to give the people our best thinking based on the research data and best practices."

*Notes:*

In 2005, in Washington, DC, 141 mayors identified environmental sustainability as a critical factor for American cities. They said, "We signed the U.S. Mayor's Climate Protection agreement[104] charging ourselves with creating implementing sustainability plans based on best global practices."

In a November 9, 2005, *Atlanta-Journal Constitution* article,[105] the journalist Cynthia Tucker wrote that Mayor Franklin had done much but was not done. She praised the mayor, stating Atlanta was experiencing a renaissance. Middle-class residents were moving in (not out), crime continued to drop, and competence had found a home at city hall after a lenghty absence. The city faces several challenges, but it is no longer struggling against a hopeless cycle of decline. Some of Atlanta's good fortune was born of

---

103    Silia Brush, "The Pipe Dreamer," http://www.usnews.com/usnews/news/articles/051031/31franklin.htm

104    http://www.usmayors.org/climateprotection/documents/mcpagreement.pdf

105    *Atlanta Journal-Constitution* archives: Cynthia Tucker, Nov 9, 2005 http://docs.newsbank.com/s/InfoWeb/aggdocs/NewsBank/10E10AC0DE29D510/0D57227E12704560?p_multi=AJBK&s_lang=en-US

sheer luck and a convergence of factors that have made urban living more attractive. The article goes on to state:

In the 2000 census, as a result, the city recorded the first increase in its white population in forty years. The housing boom has revived marginal neighborhoods, supported new restaurants and shops, and encouraged the pedestrian-friendly development that urban planners have long touted. Atlantic Station represents the nation's biggest brownfield[106] reclamation project.

Before its redevelopment as the mixed-income Centennial Place, old Techwood Homes presented a physical barrier to the northwest edge of downtown, separating Georgia Tech from the city's core. Now, the genial Centennial Place is a bridge, not a barrier, and the urban university is spreading toward the center of town.

In 2006

Mayor Franklin's second term would not only focus on infrastructure, but her agenda[107] would also include education, homelessness, information technology, and economic development.

January

- In her second inaugural speech,[108] she stated, as with her first speech, to Justice Sears, "I am honored to stand on the stage with you as you make history." Then she thanked the other dignitaries, ditinguished guests, family, and friends, saying, "I am deeply humbled by the vote of confidence the people of Atlanta have once again entrusted in me as I begin my second term as mayor of this great city. I am guided through my days

---

106    Land that has been developed for mineral extraction or waste disposal for landfill purposes and where provision for the restoration has been made through development control procedures.

107    Article; American City & County, Dec 2005, Kim O'Connell

108    http://www.atlantaga.gov/mayor/inauguralspeech2006_010306.aspx

by what Mary McCloud Bethune said: 'Faith is the first factor in a life devoted to service. Without it, nothing is possible. With it, nothing is impossible.'" America is a family. We must always help our own, especially when they are in need (speaking of the Tsunami and Katrina victims)."

She went on to state: "When I took office four years ago, I did so at a time of cynicism and anxiety for our City. Today we are standing together and looking forward, toward the future, building an Atlanta that serves as a beacon to all those communities who work together in partnership for a better future for our people. During my first term, I had four basic goals and lived their dreams. 1) Create an honest, open and caring government; 2) build a safer, cleaner city; 3) make city operations more efficient and effective; 4) repair our infrastructure by building our water ansd sewer system and expanding Hartsfield-Jackson Airport. We are making progress. Some of our goals have been reached more quickly than others, but we are making substantial progress in all these areas, as we will continue to build a government that our people deserve and expect.

These goals are the foundation of the priorities of my second term. First, focus like a laser beam on public education. Second, improve the quality of life in our neighborhoods all over Atlanta. Third, we must do more to help the homeless and those in need. Fourth, is to expand economic development and opportunity for all who make Atlanta home." In closing, she said, Ladies and gentlemen, let's blend our vision, reason and courage into a better Atlanta. We must make the tough decisions and do what we know is right, and do it with everything that we have."

- Mayor Franklin kicks off the city's infrastructure technology (IT) expo to promote the use of IT to help expand the use of technology's benefit to improving business processess, and announce that the event would be held annually.

- Mayor Franklin took a group of community business leaders from Atlanta on a business trip to China with the intent of increasing the city's economy and to encourage the Chinese to establish a consulate in Atlanta.

- The City of Atlanta began trial of new multiuse parking meters. Each "multiuse" meter will accept cash, credit, and debit cards and collect fees for seven different spaces, making parking more efficient and effective. The highly secure, all-steel machines are equipped with an alarm mechanism that sends a warning to police at the hint of vandalism, a feature which may decrease meter theft. These meters are a win-win prospect, said the city's public works commissioner, David Scott. "They will make collection of parking fees more convenient for consumers and more secure for the system. They will also allow us to reduce the amount of hardware on the curb, which will improve the aesthetics of our streets."

February

- The City of Atlanta announced "Enhance the Building Permit Process." As part of a citywide initiative to create a more effective and efficient government, the City of Atlanta's Department of Planning and Community Development had implemented major changes to improve the permitting process for Atlanta contractors and residents. On February 6, 2006, the Bureau of Buildings launched two new online services. Now, contractors and residents can view their building-permit status and find the cost of their permit online.

May

- Mayor Franklin urges communities in unincorporated South Fulton County to strongly consider Annexation into the City of Atlanta. In a May 30, 2006, letter to the president of the Atlanta City Council, Lisa Borders, and members of the city council and Mayor Franklin began by describing the city's general position regarding changes occurring in Fulton County. "Given the upcoming referenda in north and south Fulton County, I am frequently asked whether the City would be interested in annexing portions of Fulton County that are currently unincorporated (etc, etc). Of course, the City of Atlanta is enjoying a strong burst of organic growth on its own. Based on our current estimates, our population is increasing by approximately 10,000 residents each year without any boundary expansions. While this growth and the increasing prosperity of our residents can be attributed to a number of factors, we believe that our infrastructure investments, improved services and commitment to sound fiscal management have made the City a more attractive place for residents and businessess to relocate. These efforts include:

Upgrading our water/sewer infrastructure-

Lowering crime rates-

reating healthy finances-

Investing in infrastructure such as the BeltLine, parks capital improvements, and streets and sidewalks-

Advancing a variety of economic development initiatives including our New Century Economic Development Plan, our Economin Opportunity Fund, and our use of Tax Allocation Districts to accelerate development within the City(etc, etc).

As communities in unincorporated Fulton County deliberate on their future, we hope that they will strongly consider joining a healthy and prosperous City of Atlanta. We think of no municipality in the state better positioned to invest in and serve a community than the City of Atlanta.

- The fifth runway at Hartsfield-Jackson International Airport scheduled to open on time and on budget.

June

- Mayor Franklin and the chief information officer, Abe Kani, at the Muniwireless Conference in Silicon Valley, California, announcd the City of Atlanta was actively seeking a public/private partnership to deploy an affordable wireless network throughout the City of Atlanta's 132 square miles at no cost to the City. Wireless Atlanta would provide low-cost, high-speed Internet access to residents, businessess, and visitors anytime and anywhere in the city. The city would be able to use the network for mobile connectivity, to strengthen public safety, and streamline government operations. "The Wireless Atlanta initiative would address tremendous value to our citizens, businessess, and visitors with no use of taxpayer dollars", stated Abe Kani.

A request for proposal (RFP) would be uploaded on the city's webpage for interested parties. The RFP set forth the specifications of the broadband network, addressing issues including the network's business model and services, coverage area, open access policies, network infrastructure, customer services, security and privacy, as well as contributions the network provider would make regarding the other important elements of the Wireless Atlanta strategy.

- The City of Atlanta would complete the purchase of the Bellwood Quarry.

October

- Atlanta welcomed eleven Sister Cities for a Global Conference on Economic Development.

- Coca-Cola CEO announced land distribution to the City of Atlanta for the Civil Rights Museum.

November

- The US Commerce secretary Carlos M. Gutierrez announced the first Americas Competitiveness Forum[109] would be held June 11–13, 2007 in Atlanta. High level government officials from each of the hemisphere's thirty-three countries with democratically elected governments were invited to the forum to discuss how best to enhance the region's ability to compete more successfully in the global marketplace. Atlanta was chosen as the location for the forum for its commitment to competitiveness and its success in leveraging public-private partnerships to foster innovation and business development. Atlanta embraces best practices and seeks to be a best-in-class city.

- Mayor Franklin received the "Ethics Advocate Award" from the Southern Institute for Business and Professional Ethics, which recognizes an individual whose exceptional efforts are consistant with the institute's mission of promoting ethics in

---

109    City of Atlanta Online, http;//174.37.215.145/media/nr_forum_120506.aspx

business and society. Since 2003, four other people had been recognized.

December

- Mayor Franklin, Atlanta human rights dignitaries, Coco-Cola executives, Central Atlanta Progress (CAP), the Atlanta office of the Boston Consulting Group (BCG), community leaders, scholars, and citzens commenorated the groundbreaking for the Center for Civil and Human Rights. A 2.5-acre site (donated by the Coca-Cola Company) at Pemberton Place in downtown Atlanta would have a $136 million projected cost.

- The ATL READ Automated Meter Reading (AMR) began. The project would last approximately three years and the purpose was to allow the Department of Watershed Management to install a radio-based, electronic AMR system that would allow Atlanta's water meters to be read electronically. The project would include repairing and replacing Atlanta's existing residential and commercial meters and installing the AMR system.

Year in Review

At the annual Dr. Martin Luther King Jr. commemorative service[110] held on the federal holiday at Ebenezer Baptist Church, Mayor Franklin joined several distinguished speakers to honor the life and legacy of of Dr. King, and to encourage Atlanta residents to work tirelessly to fulfill Dr. King's vision of peace and equality. Living in a city that's the birth home of Dr. King and the location of the King Center, Atlanta residents, in particular, have an "obligation" to uphold Dr. King's legacy of action against inequality, poverty, violence, and discrimination in their daily lives, said Mayor Franklin.

---

110    City of Atlanta Online http://174.37.215.145/media/citynewsbytes_011706.aspx

"I am asking the people of Atlanta to do three things this year: *First*, make a commitment to help the young, the old, the needy and poor. *Secondly*, demand the federal funding needed by the victims of Hurricane Katrina and other disasters to be paid. We must stick with the victims until they reclaim their lives and dignity. And *lastly*, and most importantly, let's comprehend the full message of Dr. King."

Mayor Franklin also spent January mourning the death of Coretta Scott King with the people of Atlanta, and in her honor the mayor ordered the lowering of all the flags at City of Atlanta facilities to half-staff.

PARKatlanta.org[111] would enforce the City of Atlanta's parking meters and zones regulations. On its website it states: "Parking regulations will be enforced by trained personnel commissioned by the City to apply the law uniformly and to treat all people respectfully. Motorists can expect fair, uniform enforcement and courteous treatment and service from all enforcement and customer service personnel associated with the PARKatlanta program. Parking Zones: Parking Zones throughout the city are indicated using signs and meter decals/rate placards designating posted time restrictions."

In February, Mayor Franklin attended and spoke at Coretta Scott King's "Celebration of Life" homegoing service.

In February, Mayor Franklin was profiled on a C-SPAN Sunday broadcast.

In March, the City of Atlanta hosted its 2006 Women's History Month Celebration.

Mayor Franklin travelled with CARE[112] to South Africa (the trip would offer women in the United States an opportunity to stand in solidarity with women in poor countries). CARE is a leading humanitarian organization

---

111    http://www.parkatlanta.org/regulations.html

112    http://www.care.org/about/index.asp

fighting global poverty. Women are the heart of the community-based efforts to improve basic education, prevent the spread of disease, increase access to clean water and sanitation, expand economic opportunity, and protect natural resources.

The City of Atlanta sue online travel sites for millions in unpaid hotel room taxes.

In April, the City of Atlanta launched the Integrity Matters program. Mayor Franklin and the Chamber of Commerce joined forces to find summer internships for students.

The City of Atlanta launched the first citywide reading initiative, Atlanta Reads: One Book, One Community, with *Run With the Horsemen* selected as the inaugural text.

In May, Mayor Franklin vetoed the city council's ethics legislation.

In June, Mayor Franklin and the City of Atlanta, in partnership with Zoo Atlanta, offered free admisssion to city residents.

In June, Mayor Franklin was elected presdient of the Georgia Municipal Aassociation.

In June, Mayor Franklin provided the keynote address at the luncheon honoring the US finalists for the Stockhlom Junior Water Prize (the most prestigious international high school competition related to water-science research) at the downtown Sheraton Hotel. The Stockholm Junior Water prize is open to projects aimed at improving the quality of life through improvement of water quality, water resources management, water protection, or water and wastewater treatment. The US winner advanced to the international competition in Stockholm, Sweden, during World Water Week.

On June 23, a thirty-two-million-dollar deal was announced that would prevent the Martin Luther King Jr. papers from being auctioned off and keep them in Atlanta.[113] A private coalition of business and civic leaders bought the collection from the King family following concerted efforts by Mayor Franklin and the former mayor Andrew Young. Under the deal, Morehouse College, King's alma mater, will own the collection.

In July, Mayor Franklin's office and councilman C.T. Martin presented the sixteenth annual Youthfest 2006 at Wilson Mill Park (Adamsville).

In August, the Chamber of Commerce and business leaders head to China. This would lead to the Chamber of Commerce mission, led by Mayor Franklin, to expand Atlanta's reach around the world.

In October, the president of the Atlanta Development Authrotity gave his letter of resignation to Mayor Franklin.

In November, the City of Atlanta welcomed forty-five international women leaders for the first workshop on gender Equality. Atlanta hosts the third Mayors Against Illegal Guns coalition regional conference. Senior officials from thirteen municipalities, representing five states in the Southeast, discuss the national fight against illegal guns.

*Note:*

Atlanta is home to not only the world's busiest airport but also the most efficient.[114]

Over eighty-five million passengers pass through Hartsfield-Jackson International Airport each year. It is the number one reason businesses relocate to Atlanta and is a major economic engine for the Atlanta region. There are over 105,000 airport-related jobs in metro Atlanta. It was the city's goal to increase this number to 129,000 jobs by 2009. In fact,

---

113    City of Atlanta Online http://174.37.215.145/media/nr_kingpapers_062606.aspx

114    http://www.investatlanta.com/atlCompAdvan/transportationHub.jsp

passenger and air-cargo activity at Hartsfield-Jackson generates $18.8 billion of business revenue annually to firms providing services at the airport and to local visitor-industry firms.

## Critics

Critics, the public citizens, out-of-towners, and occassional commuters had a feud feast with the mayor, Public Works commissioner David Scott, the city council, the city police, PARKatlanta, Central Atlanta Progress (CAP), and anyone else involved in instituting such a "stupid idea" as the new parking meters that had been installed and were being enforced by tickets that were then sent to motorists in the mail. *First,* putting the system in place cost the city too much for the contract. *Secondly,* when you parked your car, you didn't know where to put the money in the meter since no one knew. You had to be told to find the number painted on the curb and use that to pay at the meter. *Lastly,* people felt that PARKatlanta meter readers were stalking streets constantly, and lurking in the shadows waiting for meters to expire or to see if anyone would pay into the meter system.

In 2007

January

- The City of Atlanta Phoenix II Park named 2006 Park of the Year.

- The City of Atlanta selects Earthlink to build out its municipal wireless network

- Piedmont Road Sewer Capacity Improvement Project began on the January 29 and was expected to be a month-long project.

*Note:*

The existing ten-inch sewer pipe that runs down Piedmont Road from the Lenox Connector/Mathieson Drive intersection to within 670 feet of Peachtree Road would be replaced with an upsized fifteen-inch pipe that would be extended south toward Peachtree Road. The larger pipe was necessary to provide sewer capacity growth in the area and eliminate sewer overflows. The purpose was to use the pipe-bursting repair technique (trenchless technology) to replace approximately 1,000 lf of existing sewer line, and install approximately 125 lf of new 15-inch pipe, with six new manholes.

March

- Custer CSO Storage and Dechlorination facility is completed. Originally designed as an above-ground structure, it was redesigned as an underground storage cavern. With a storage capacity of ten million gallons, along with the thirty-four million gallons of the existing Intrenchment Creek Storage Tunnel, it gave the city an overflow storage capacity of forty-four million gallons. The dechlorination system would allow the plant to take chlorine, which is part of the treatment process, out of the flow before the treated water is discharged, thus protecting fish and wildlife in and near the receiving stream.

- The City of Atlanta Automated Meter Reading project began.

- City of Atlanta dedicates its first Boundless Playground.[115] Coan Park was selected as the site for the playground because surveys showed that the largest number of youth with special needs lived in the Coan community and surrounding neighborhoods.

- Mayor Franklin joined the task force to unveil a new vision for Peachtree Street.

---

115    http://video.google.com/videoplay?docid=-7071877449317608400

April

- Mayor Franklin and council members celebrate the groundbreaking ceremony for the Howell Mill Road widening project.

May

- New watering restrictions are set in place for the City of Atlanta due to drought conditions.

- Mayor Franklin and council members celebrate the groundbreaking ceremony for the Pryor Road Streetscape Improvement Project.

- Initiative for Georgia Tech students and faculty to conduct greenhouse gas emissions inventory for the City of Atlanta to establish a baseline to measure future emission reduction for the future. This inventory was crucial for Atlanta to assess gases that were emitted by electricity, natural gases, vehicles, refrigerants from HVACs, landfill waste, and wastewater-treatment facilities that produce methane and nitrous oxide into the atmosphere.

The table shows that electricity use rose slightly from 2005 to 2006 and was basically constant from 2006 to 2007.

- Groundbreaking ceremony on a joint police and fire headquartes in downtown that would be a best-in-class public safety and judicial services center. The new facilty, which was built to LEED Silver standards, serves as a joint administrative hub for the Atlanta Police Department (APD) and the Atlanta Fire Rescue (AFR).

| Source | Net Greenhouse Gas Emissions (thousand metric tons $CO_2e$) | | |
|---|---|---|---|
| | 2005 | 2006 | 2007 |
| Facility Electricity | 220 | 240 | 240 |
| Facility Natural Gas | ---- | --- | 36 |
| AATC Electricity | 130 | 140 | 150 |
| AATC Natural Gas | 8.9 | 8.1 | 9.1 |
| Outdoor Lighting | --- | --- | 31 |
| Traffic Signals | 8.5 | 8.5 | 8.2 |
| Transportation Fuels | --- | --- | 28 |
| Wastewater Treatment | --- | --- | 5.4 |
| Refrigerants | --- | --- | 2.9 |
| Landfill Waste Transport | --- | --- | 0.06 |
| Landfill Gas | --- | --- | 29 |
| Total | | | 540 |

- The Georgia Department of Transportation begun the Fourteenth Street Bridge improvement project, starting with utilities locations. The project was designed to relieve congestion and improve safety at the Fourteenth Street interchange and was expected to reopen in the fall of 2010.

July

- Mayor Franklin, council members, and community leaders unveil Streetscape improvements on Hollywood Road.

August

- Mayor Franklin and council members celebrate the groundbreaking ceremony for the Niskey Lake pavement improvement project.

September

- Mayor Franklin, council members, commissioner Dianne Harnell Cohen, the CEO of AT&T Mobility and Consumer

Markets, Ralph de la Vega, community leaders, and family dedicate the state of Georgia's first Hispanic Memorial Park (a tribute to Sara J. Gonzalez during National Hispanic Heritage Month)

- Fort McPherson land reuse plan[116] slated for final approval.

With state and local officials, led by governor Sonny Perdue, Mayor Franklin, Joseph Macon, the mayor of East Point, and Fulton County Commission chairman John Eaves outline economic-impact plans for future reuse of Fort McPherson, located in southwest Atlanta, to turn the property into a scientific research park after the base's closing (following the federal government's Defense Base Realignment and Closure Act of 2005) and for the McPherson Planning Local Redevelopment Authority (MPLRA). If approved by the Army, the research park would be developed as an urban mixed-use work/live/play science and research park for interdisciplinary research and technology transfer in partnership with Georgia research institutions and private partners.

*Note:*

Fort McPherson was being closed[117] under the federal government's Defense Base Realignment and Closure Act (BRAC) of 2005. Effective November 9, 2005, the Base Realignment and Closure Commission voted to close Fort McPherson, the 487-acre military base. Fort McPherson would be vacated by the US Army in 2011.

---

116    http://www.ijhssnet.com/journals/Vol._1_No._3;_March_2011/1.

117    http://www.georgia.gov/00/press/detail/0,2668,78006749_96092832_96284582,00.html

October

- The Atlanta City Council reauthorized Montgomery-Watson/ Khafra (MWH) for a seventh year under its program-management agreement on behalf of the Department of Watershed Management. MWH has been with the program since its inception in 2001.

- Mayor Franklin and council members celebrated the groundbreaking ceremony for the Simpson Road Streetscape improvement project.

- City of Atlanta dedicated new cultural center in South Bend Park.

- Mayor Franklin released new water bill usage.

November

- Mayor Franklin announced kick-off and press event for the City of Atlanta's first Comprehensive Transportation Plan (CTP).[118]

- Mayor Franklin announced a news conference to celebrate the grand opening of the new Government Center parking facility for City of Atlanta employees and the public located at 270 Central Avenue at Washington Street.

December

- Mayor Franklin and council members celebrated the ribbon-cutting ceremony for the Niskey Lake Road unpaved street project.

- City of Atlanta, the Ron Clark Academy, and HGTV participated in a nationwide contest. The campaign is called "Change the World. Start at Home." Cities from all over the nation

---

118    http://web.atlantaga.gov/connectatlanta

submitted community projects that they wanted to see happen in their city.

- HGTV narrowed it down to ten projects in ten different cities. Atlanta was chosen for its revitalization project around the Ron Clark Academy in southeast Atlanta.

- The City of Atlanta hosted the inaugural Americas Competitiveness Forum.

*Note:*

By the end of 2007, the course for infrastructure progress in Atlanta appeared so good that the results were starting to show, so much so that Mayor Franklin landed on the cover of the *Georgia Trend* magazine[119] with a painted color portrait of herself with her signature flower on her lapel. In the article labeling her a pragmatic visionary, Mayor Franklin is described as a no nonsense administrator, skilled negotiator, vigilant steward, and above all as having what makes her succeed—shear willpower.

McDaniel Street CSO and Greensferry CSO facilities were eliminated in 2007 upon completion of the sewer-separation project.

Stockade CSO was eliminated in 2007 when the additional underground storage facilities were completed at Intrenchment Creek.

Year in Review

Why is the Americas Competitiveness Forum (ACF) important to Atlanta?[120]

The forum's website states: "The forum is an incredible opportunity to elevate Atlanta's stature as a center for global business. Public officials and

---

119    *Georgia Trend* magazine, 2007

120    http://metroatlantachamber.com/content/Article.aspx?Id=168&AspxAutoDetectCookieSupport=1

business executives from more than 30 countries will attend the conference in 2010."

Mayor Franklin stated:

1) "We have the perfect opportunity to showcase why Atlanta is such a great location for global business. At the same time, our local business leaders have the chance to meet with potential customers and government officials, representing new market opportunities for local companies. It is not every day that such rich opportunities for business present themselves in our town, so we must take full advantage of the Americas Competitiveness Forum."

2) What differentiates Atlanta as a host of ACF?

The US Department of Commerce initially selected Atlanta for the inaugural forum because of our strenght in public-private partnerships. More importantly, though, the Department of Commerce has decided to bring the forum back to Atlanta for 2008 and 2010 because of these partnerships. Our community has demonstrated time and again how the public and private sectors can come together to help Atlanta thrive. The ACF is a perfect example of a partnership between the city, Metro Atlanta Chamber of Commerce, the Georgia Department of Economic Development, and CIFAL[121] Atlanta. We've established a practice of successful partnering that other cities and even countries can emulate. These partnerships are not only a point of pride for Atlanta but also a competitive advantage that will differentiate us and position us ahead of other cities as we compete in the global economy, stated Mayor Franklin.

---

121    http://www.cifalatlanta.org/about/background/

In an article[122] in *Georgia Trend* magazine in January naming Mayor Franklin 2007 Georgian Of The Year, writer Vincent Coppola praises the mayor for her fortitude in being Atlanta's leader. In the article is a photo of the mayor with this caption beneath:

From the ashes: Shirley Franklin has led Atlanta's phoenix-like rise from the depths of bankruptcy and corruption to the heights of solvency, a positive business environment and hope for the future. He writes that Franklin used the term herself in citing the example set by Coretta Scott King and other civil rights leaders who found the resolve to push, against all odds, to right wrongs and drive reform. In five short years Mayor Shirley Franklin has seemingly reversed the course of a city hamstrung by corruption and mismanagement, a city bleeding red ink as surely as its antiquated sewers were bleeding wastes, where toxic fumes of cynicism and discouragement afflicting so many urban centers had begun to seep. With an unlikely combination of can-do enthusiasm and rigorous discipline, Franklin has rekindled the sense of promise, rebirth and reconciliation that defines the best of Atlanta. "I could not imagine being mayor of a city that was broke. The job is hard enough if you pay your bills, but if you're running from your creditors and can't spend money to hire police officers and upgrade services, it's unthinkable. I had this great fear of coming to work every day trying to hold the place together with hairpins, so the decisions that would put us on a fiscally sound approach was relatively easy to make."

The state of Georgia, stricken by months of drought, confirmed that it would sue the Army Corps of Engineers. The governor, Sonny Perdue, had said in October the state would seek an injunction forcing the Corps of Engineers to stem the flow of water from Lake Lanier, Atlanta's primary

---

122    http://www.georgiatrend/January-2007/2007-Georgian-of-the-Year

water source.[123] The Corps administers the lake, which supplies most of the water to Georgia's capital and feeds the Chattahoochee River, which winds through three states. Rainfall in the area was about fifteen inches below normal for the year. Mayor Franklin said, "This is dire, severe, extreme drought." In the City of Atlanta and surrounding counties, outdoor watering was banned except for a few commercial uses. The Army Corps of Engineers said there was about a three-month supply of water left in Lake Lanier, which was fifteen feet below its capacity.

In May 2007, the Federal Aviation Administration released a report titled "Capacity Needs in the National Airspace System, 2007–2025" as part of its Future Airport Capacity Task (FACT2).[124] The report identified Hartsfield-Jackson International Airport and the Atlanta metropolitan area as needing additional capacity by 2025. Following the report's release, Atlanta was given a one-million-dollar federal grant to study the possibility of adding another airport.

In July 2007, the Atlanta Strategic Action Plan (ASAP) was set for round two in a series of meeetings with the city to address Atlanta's becoming a world-class city. The agenda would focus on elements in and around the city to include population, transportation, preservation, diversity, economic development, housing, parks, recreation, open space, community facilities, neighborhood planning units (NPU) land use, and the next steps to be developed to move forward. The community meetings also addressed the findings and recommendations for each element discussed.

---

123    http://edition.cnn.com/2007/US/101/18/pip.atlantadrought, "Drought-stricken Georgia says it will sue over water," October 19, 2007.

124    www.wikipedia.com; jump to: navigation, search- Atlanta's second airport.

*Note:*

With the addition of the fifth runway, that should have relieved capacity needs at Hartsfield-Jackson International Airport from the FACT1 and FACT2 results of the 2003 comparison of airport needs.

Both Delta Air Lines and AirTran Airways, which both operated their primary hubs in Atlanta, expressed skepticism about building a second airport.

In 2008

January

- Governor Sonny Perdue delivered his annual State of the State Address to the General Assembly. In outlining his priorities for 2008, some of the items on his agenda were a proposal to fund a transportation infrastructure bank and plans for water infrastructure, and reservoirs.

- The commissioner of watershed management, Rob Hunter, the commissioner of Parks and Recreation, Dianne Harnell Cohen, and the chief of staff, Greg Pridgeon, held a press conference to announce the City of Atlanta's policy for outdoor events in parks in response to Level 4 drought restrictions.

- Mayor Franklin went to the city council with bad news. The city had a projected budget gap of seventy million dollars, and she attributed the deficit to a slowing economy, faulty estimates, and growing pension and health insurance costs.

- Mayor Franklin and the city council authorized the city to participate in a one-million-dollar toilet rebate program to be administered by the Metropolitan North Georgia Water Planning District. "Toilets are the largest water user and water wasters in the average home," stated Commissioner Rob Hunter.

"Through this program we hope to encourage our customers to replace their inefficient toilets with new water-saving models. Doing so will not only lower their water/sewer bills, it will also help Atlanta conserve water in this serious, ongoing drought."

February

- Sustainable Atlanta initiative began.

  Lead, Change, Sustain.[125] Its website states: "Sustainable Atlanta was created to capture the mayor's long-term view. Sustainability addresses multiple forms of health—human, financial and environmental. Achieving those protections is accomplished by governments, corporations, businesses and residents taking action at all levels, from installing compact fluorescent light bulbs that immediately save energy to creating sustainability policies that position us better for tomorrow."

March

- Georgia DOT Downtown Connector resurfacing project began.

  At a cost of $27.7 million, the purpose of the project was to resurface a portion of the I-75/85 Downtown Connector corridor. The roadway was last resurfaced in 1995, prior to the 1996 Summer Olympics. The total length to be resurfaced was 4.79 miles, between University Avenue and Tenth Street in the northbound and southbound lanes.

May

- The commissioner of Parks and Recreation, Dianne Harnell Cohen, the director of the Office of Recreation, Charlene

---

125    http://www.sustainableatlanta.org/report/Sustainability%20Report.pdf

Braud, and council members celebrated the renaming of the Piedmont Park Tennis Center in honor of the former Tennis Center facility manager Sharon E. Lester.

- Mayor Franklin, Fire department chief Kelvin Cochran, and Atlanta firefighters celebrated the groundbreaking for the new fire station number 13. The new fire station would replace the station located at 447 Flat Shoals Avenue.

- The Department of Public Works and the Georgia Department of Transportation announced the closure of the Mitchell Street Bridge, effective immediately. Mitchell Street, between Northside Drive and Spring Street, closed to all traffic. An inspection of the bridge revealed that it needed to replaced and had been deemed unsafe for vehicular traffic.

- Techwood Drive/Fourteenth Street ramp closed. This closure was a necessary part of the ongoing project to replace the Fourteenth Street Bridge over the downtown connector, relieving congestion at the Fourteenth Street interchange. The ramp's reopening was scheduled for the winter 2009.

September

- Mayor Franklin and the citizens of Atlanta remembered September 11. The City of Atlanta also wants to acknowledge the fearless commitment of its public safety employees, who are first responders every day and whose courage reminds us all of the ultimate service to others.

- City of Atlanta's Connect Atlanta team consisting of city planners, city council members, technical and stakeholder advisory committees, NPU and neighborhood leaders, and the general public developed its first master transportation plan.

October

- Mayor Franklin and council members celebrated the ground-breaking for the Historic Fourth Ward Park, the thirty-five acres of blighted industrial lowland across North Avenue from City Hall East. It was the first park in the emerald necklace that would ultimately be Atlanta's BeltLine.

November

- City of Atlanta's Departments of Planning and Community Development, Watershed Management, Parks, Recreation and Cultural Affairs, Public Works, the Atlanta Police and Fire department, in observance of Geographic Information System (GIS) Day 2008, unveiled its public GIS website. The site provides an array of geographic information, including a map catalog, interactive maps, and property information such as zoning, future land use, flood hazard, tax records, and permit history.

- The Office of Sustainability and the Office of Enterprise Asset Management announced they would work with departments across city government to improve current programs and policies and implement new ones that would not only be good for the environment but also save the city money over time, said Mayor Franklin. The collaboration's aim was focused on the implementation of no-cost energy and water-conservation measures that would reduce both utility costs and greenhouse gas emissions in City Hall. This was a result of the reprogramming of heating and air-conditioning controls to ensure that systems shut down as much as possible during unoccupied periods, thus resulting in an expected 15–20 percent energy savings and an annual cost savings of over $120,000 for the City Hall building alone. Other cost saving efforts included:

Power to Change Campaign

Utility Management System

Green Revolving Loan Fund

Carbon reduction efforts

Teleworking and Compressed Work Week

December

- KPMG conducted a performance review audit of the Department of Watershed Management from the passed city council resolution 08-R-1014 of May 19, 2008 requesting the audit in conjunction with considering the department's proposed water and sewer rates. KPMG conducted its work from December 2008 through April 2009. The methods of the audit included: 1) interviewing Department of Watershed Management staff and consultants and other city staff; 2) reviewing previous audit work and other studies; 3) reviewing and analyzing operational and financial data, including department policies and procedures manuals, organization charts, DWM's strategic plan, interjurisdictional agreements, fee schedules, monthly aged receivable reports, bond rating reports, revenue reports by type of customer, fund account structure and current balances, billing adjustments, budget documents, audited financial statements, and the city's cost allocation plan; 4) reviewing the capital improvement process, including budgeting, scheduling, design, procurement, contract administration, payment applications, value engineering and evaluation, change orders, use of contingencies, retainage, and materials management; 5) selecting a judgmental sample of nine construction projects for detailed file review to assess compliance with processes; 6) surveying employees to ask about perceived departmental strengths and areas for improvement 7) reviewing rate model revenue and expenditure forecasting and assumptions and analyzing the effect of changing certain assumptions to better reflect historical data and changed market conditions.

- The City of Atlanta adopted a new rate program.

- The City of Atlanta launched a new website, SustainableAtlanta. org, to encourage community involvement and education on water conservation, recycling and materials management, green building, and the green industry.

- Atlanta City Hall sidewalk/parks improvement project began.

Year in Review

Mayor Franklin spent the year focused on many issues, including infrastructure, energy efficiency, and connecting youth to jobs and positive responses to immigration, a public-private partnership that will serve thousands of Atlanta students with educational opportunities, scholarships, employment training, and career assistance. "Atlanta has made great progress implementing effective programs that ensure that young people have the best opportunity to succeed. I am anxious to learn how we can contribute to improve and expand on what we're doing in Atlanta," said Mayor Franklin at the Mayors Innovation Project meeting at the Brookings Institution in Washington, DC.

In May, Mayor Franklin forwarded her fiscal year 2009 proposed budget to  council members for review to include in the budget an overall assessment of the economic and fiscal health of the city. She laid out four Principles behind the budget:

1) Transparency

2) Reform

3) Public Safety

4) Operational Efficiency

In June, Mayor Franklin was awaiting a response on the prosposed budget from the president of the city council, Lisa Borders. She sent her a letter stating, "I understand that you have reviewed two significant proposals to increase revenues and decrease costs: the possible re-amortization of the pension plans and the sale of certain property tax account receivables." She went on to state, "I have been informed by the Chief Financial Officer that we are beginning to see downward pressure on our revenues that could well impact overall revenue collections next year and give us a margin for error." Mayor Franklin goes on further in her letter to explain all the points outlined in her proposal and states, "I remain convinced that the Administration's proposal represents the best way to balance the FY09 budget." Mayor Franklin ended the letter by stating, "For all these reasons I encourage you to consider the proposed budget and to vote on it within the timeframe imposed by the City Code and State law."

A severe drought that had begun in 2007 and ended earlier in the year prompted Atlanta to take serious steps to further reduce water use. The city declared Level 4 restrictions—the strongest—several months before the state implemented them. Officials also created a number of conservation programs and distributed water-conservation kits, flush valves and "instant-off" devices for faucets. They offered free water audits; rain barrels construction programs; educational workshops for residents, landscapers, and large users; toilet rebates; and new-toilet installations for low-income and elderly customers. They established the Save Water Atlanta Team to enforce watering restrictions. The city had already put in place a three-tiered conservation rate structure that rewards low use.

Mayor Franklin, SCANA Energy, the Office of Constituent Services, and the Atlanta Fire Rescue team up for the seventh consecutive year in the "Help Atlanta's Elderly Stay Cool Campaign," an effort to ensure that Atlanta's elderly have a safe and cool summer, including distributing one thousand fans to seniors throughout the communities.

The City of Atlanta authorized a one-million-dollar toilet-rebate program.

Atlanta City Council approved water-sewer rate hikes.

Mayor Franklin was named by The White House Project as one of "8 in '08." She was the only person on the list who wasn't a governor, a senator, or a presidential cabinet member.

Atlanta's chief financial officer (CFO), Janice Davis, surprisingly turned in a brief letter of resignation[126] in July to the Atlanta City Council.

Atlantan's remembered September 11.

City of Atlanta wins a $300,000 tree planting grant as part of the American Express Root for Our City Challenge, an initiative designed to help make eight US cities cleaner, greener, and more beautiful. The tree-planting program would plant 900 trees and maintain them in residential neighborhoods and retail districts throughout the city.

Facing a major deficit, Mayor Franklin asked the Atlanta City Council to approve a property tax increase to avoid cuts in public safety and was unanimously shot down. In December, Mayor Franklin announced 222 city workers would loose their jobs to help fill a projected budget shortfall of fifty to sixty million dollars.

The Center for Civil and Human Rights announced its official site at Pemberton Place, adjacent to the World of Coca-Cola, the Georgia Aquarium, and Centennial Olympic Park. The Freelon Group (Phil Freelon),

---

126    http://www.11alive.com/news/local/story.aspx?storyid=116882&catid=3

an architectural firm in Durham, North Carolina, would partnered with HOK, an architectural firm in Atlanta, to design the center. The planned LEED-certified center will be designed to serve as a museum and cultural research center portal for archives, education, discussions through performances, lectures, symposia, and partnerships across Atlanta, the state, and the nation. The center would cover twenty thousand square feet and cost an estimated $125 million. H.J. Russell & Co., C.D. Moody Construction Co., and Holder Construction Co., three Atlanta-based construction firms, teamed up in a joint venture to provide construction-management services to build the facility. Groundbreaking was scheduled for the summer of 2012, and opening in 2014.

The US Department of Housing and Urban Development (HUD) Neighborhood Stabilization Program gave the City of Atlanta $12.3 million (through the HOPE program) in an effort to stabilize communities devastated and hit hardest by foreclosure in Adair Park, Capital View, English Avenue, Mechanicsville, Pittsburg, Summerhill, Slyvan Hills, Vine City, Washington Heights, and West End.

Listed below are Atlanta's positions in the 2008 SustainLane Rankings.[127]

3rd Green (LEED) Building

11th Local Food & Agriculture

11th Metro Transit Ridership

17th City Commuting

17th Water Supply

18th Natural Disaster Risk

18th Energy & Climate Change Policy

21st Green Economy

---

127    http://atlantasustainabilityweek.org/ATLSustainPlan.pdf

22nd City Innovation

29th Solid Waste Diversion

34th Housing Affordability

37th Knowledge Base & Communications

40th Water Quality

42nd Air Quality

42nd Planning & Land Use

45th Metro Street Congestion

❖ Rankings out of 50.

In August, Atlanta City Council president Lisa Borders, announced she was dropping out of the race for mayor.[128] She cited personal family issues and her priority in her current role as council president. The still-large field of prospective candidates included councilman Ceasar Mitchell, councilwoman Mary Norwood, and state senator Kasim Reed.

*Note:* Mayor Franklin was nearing the end of her second term as mayor and cannot run again.

In December, Mayor Franklin sent a letter (basically an SOS) to Georgia congressman John Lewis to inform him of the current state of the city and three factors that were hampering progress and growth:

1) Costs to service existing debt are going up.

2) Increasing costs of new debt.

3) Threats to operating revenue.

She stated, "The purpose of this letter is to describe for you how the deteriorating financial climate is directly impacting the city's two major infrastructure programs—the Clean Water Atlanta water/sewer program and

---

128    http://www.11alive.com/news/local/story.aspx?story=119685

the expansion program at Hartsfield-Jackson International Airport." She continued by saying, "These infrastructure projects support hundreds of families and many businesess. Investing in our infrastructure will create immediate jobs in the city." With revenues already falling because of conservation efforts in response to the regional drought, the Department of Watershed Management (DWM) entered the economic downturn in a vulnerable financial position. Mayor Franklin was explaining in closing that *without access to capital*, other construction projects in jeopardy included: new cargo facility construction, a runway extension project, concourse expansion, airport fire station renovation, and phase four of the terminal upgrade project at the airport. In conclusion, Mayor Franklin would state, "We urge you and your colleagues to seriously consider taking actions that will specifically address the threats the city faces" that might include:

1) The creation of a short-term borrowing facility that would allow cities to meet immediate cash-flow needs.

2) Direct intervention in the credit markets—either through capital inflows or credit gaurantees—that will increase the demand for municipal debt and lower its costs to issuers.

3) Direct infrastructure investments (either through a national infrastructure bank or through direct grants and loans) that will permit cities to continue their infrastructure programs while the credit markets stabilize and recover.

## Critics

Mayor Franklin, in an ongoing dispute with the council members, raged throughout the year because the council adopted a 2009 budget in June that slashed hundreds of city jobs to offset a projected $140 million shortfall.

Praise

US District Court judge Thomas Thrash overseeing the compliance with the Consent Decrees, stated in a 2008 status hearing that the projects' having been completed on time and on budget was nothing short of amazing. "Frankly, I expected excuses, delays, obstructions, and incompetence, and under Mayor Franklin's administration, none of that's happened. The work's been done. It's been done on time, and I think pretty much done within budget. And it really is a remarkable accomplishment.[129]

In 2009

January

- In Mayor Franklin's final State of the City address to the city council, she outlined in a fifteen-minute speech a list of accomplishments that she said had put the city on "the threshold of greatness." She also stated that "as much as we have accomplished, we have many more mountains to climb—we cannot wait till 2030."

How good is Mayor Franklin?[130] The website of City Mayors invited its readers to go online and assess the performance in office of Shirley Franklin. Readers were asked to assess and rate the overall performance on a scale of 1(extremely poor) to 10 (outstanding). The website provided tracking of the mayor's performance to be published and displayed. The following results were published from a performance index.

April 2009: 6.08 points out of 10

June 2009: 6.12 points out of 10

November 2009: 5.85 points out of 10

---

129    http://fiscalresearch.gsu/atlanta case study/Watershed%20Mgmt%20Case20Study.pdf

130    Article: www.citymayors.com, How good is Mayor Shirley Franklin? Mayor of Atlanta, USA

February

- Congress approved and President Obama signed the American Recovery and Reinvestment Act (ARRA).

- Mayor Franklin, commissioners, and executive director of Atlanta's Workforce Development Agency, Deborah Lum, held a news conference to discuss President Obama's Stimulus plan and how it would affect Atlanta as well as Georgia.

- Mayor Franklin held a press conference to discuss the state of the city. She said that "over the last seven years, we have been hard at work in the remaking of Atlanta and the Obama administration promises a great lift to our collective efforts . The urban policy advanded by the Obama administration will support and fund programs critical to the long-term success of Atlanta. This includes more federal funding for our roads, bridges, and sidewalks and our region's largest transit operator MARTA. Obama's focus on sustainability and emphasis on reducing greenhouse gas emissions is in synch with our own sustainability initiatives and our desire to locate more of our citizens to transit, with the construction of the BeltLine and the Peachtree Streetcar." Mayor Franklin goes on to state, "I am excited about the personal connections we have in the Obama aministration to help us forge a close working patnership. Atlanta shares Obama's vision for the future"

- Mayor Franklin, along with a delegation of the US Conference of American Mayors, met with White House senior advisors regarding the American Recovery and Reinvestment Act. The purpose was for urban city mayors to stress that investing in Main Street metropolitan economies, which comprise 90 percent of cities' gross domestic product and drive the economy, is the most direct path to creating jobs and stimulating business that can begin to reverse the current economic downturn. Mayors believed that the American people could not afford delay and were urging Congress to quickly pass a recovery billnow.

- Governor Perdue, after first cautioning against accepting federal stimulus funds from the American Recovery and Reinvestment Act (ARRA), approved the use of the funding to pay for two infrastructure projects in the state.[131] Governor Perdue said "investment in our state's water and sewer infrastructure stimulates the economy."

March

- Mayor Franklin hosted the American 2050 forum[132] at Georgia Tech's Global Learning Center[133] in Atlanta, along with these sponsors: Regional Plan Association, Center for Quality Growth and Regional Development, Atlanta Regional Commission, Georgia Transportation Institute , Metro Atlanta Chamber of Commerece, Livable Communities Coalition, The Medical Center, North Georgia RDC. The focus was on the "Piedmont Atlantic Megaregion (PAM) in the Global Economy," which is a national initiative to develop a framework for America's future growth and development in face of rapid population, demographic change, and infrastructure needs in the twenty-first century. The focus of the forum was to bring together business leaders, civic leaders, academic thinkers, and policy makers to discuss transportation and infrastructure issues in their regions,

---

131    http://politicalcorrection.org/blog/200910210002

132

133    http://www.cqgrd.gatech.edu/proceedings/pam_forum_2009/index.php

stretching from Raleigh, North Carolina, to Birmingham, Alabama.

- The City of Atlanta joined 540 cities in seventy-five countries to show support for the World Wildlife Fund (WWF), the largest global climate event in history. The City of Atlanta agreed to "turn out"as a flagship city for Earth Hour 2009. On Saturday, March 28, 2009, from 8:30 p.m.–9:30 p.m. local time, individuals, businessess, and local government officials throughout Atlanta were encouraged to take part in this historic evening by turning off nonessential lights for one hour in support for a call to action on climate change in the United States and around the world. Hundreds of major landmark buildings in Downtown, Midtown, and Buckhead as well as many neighborhoods throughout the city, were anticipated to go dark in honor of the event. This is the second year Atlanta would be an official flagship city.

- The City of Atlanta announced the municipal carbon footprint for government offices. The city will announce the total greenhouse gas emissions for the Atlanta city government. The carbon footprint report established a baseline for Atlanta's government, which aimed to reduce emissions by 7 percentby 2012 (see charts ).

- The City of Atlanta's Department of Parks, Recreation & Cultural Affairs was awarded accreditation status. This national recognition by the National Recreation and Parks Association is the highest honor that can be bestowed on a parks, recreation, and cultural affairs agencies and sets them apart from thousands of other agencies and systems throughout the nation, said Dianne Harnell Cohen, the commissioner Parks.

- City of Atlanta launched the ATLStat[134] public website, a performance management system of municipal operations. ATLStat provided Mayor Franklin and her senior management team with real-time information regarding the performance and effectiveness of city services. The ATLStat process would have three primary goals: 1) information- to provide periodic operating statistics to the mayor and chief operating officer that convey the quality of service delivery; 2) management- to provide a management tool that increases accountability in senior management by setting performance targets and tracking progress; 3) and transparency- to create a public window into the performance of city services.

April

- Mayor Franklin was awarded the Urban Leadership Forum Award by the University of Pennsylvania's Institute for Urban Research. The honor is given to public servants who have excelled in a particular area of urban leadership. The year focused on building competitive and sustainable cities, and Mayor Franklin was one of three honorees.

- Mayor Franklin, Home Depot volunteers, KaBoom, and the Progressive Redevelopment Inc. joined together to build a state-of-the-art, 2,500-square-foot playground in one day at the Seven Courts Apartments, off Martin Luther King Jr. Drive SW.

- The City of Atlanta's Weed and Seed program and seven Atlanta neighborhoods helped to beautify communities in Mechanicsville, Pittsburg, Vine City, English Avenue,

---

134    City of Atlanta Online http://174.37.215.145/media/nr_atlstat_031009.aspx

Summerhill, Peoplestown, and Adair Park. The goal was to partner in beautifying neighborhoods in an effort to promote safety, to reclaim their communities, and encourage neighborhood pride.

- Mayor Franklin, city council members, and a host of local companies hosted the largest single-day electronic-waste collections in the southeast on Saturday April 18 from 9 a.m. to 1 p.m. at Turner Field's Orange Lot.

City of Atlanta held a one-day electronic-waste, tires, batteries, recycling collection, copiers, printers, CD players, VCRs, telephones, computer monitors, and microwave ovens at Turner Field. Under Mayor Franklin's leadership, Atlanta was more committed to preserving the environment than ever before, stated the mayor. The goal of Electronic Waste Recycling Day is to promote the recovery, reuse, and recycling of obsolete electronic equipment that would usuly go to the landfill.

May

- Oakland Cemetery receives May's Park of the Month honors.

- Mayor Franklin had to attend the tri-State water hearing at Bryan Simpson US Courthouse in Jacksonville, Florida, relating to a number of water lawsuits concerning the Georgia-Florida boundary battle, and on issues of eminent drought at Georgia's lakes.

- Mayor Franklin and the chief operating officer, Greg Giornelli, presented the $541 million budget for fiscal year 2009 Presentation to the Atlanta Committee for Progress. The presentation highlighted the measures and reforms recommended by the Franklin administration. The group was also briefed on the status of the city's water/sewer infrastructure improvement program and the proposed rate increases necessary to continue the program. The budget plan addressed four ways to bridge a budget gap, with explanations, charts, and graphs:

1. Spending cuts

2. Revenue initiatives

3. Revenue anticipation rate

4. Property taxes

The chairman of the Atlanta Committee for Progress (ACP), Neville Isdell, commended the mayor for facing the challenges inherent in the budget process and making the tough decisions necessary, stating, "We strongly support Mayor Franklin's balanced approach to financing government, reducing costs and increasing revenue, all the while ensuring that critical matters such as public safety are not compromised. The ACP unanimously endorsed the FY09 budget proposal and water/sewer rate increases." The commissioner of Watershed Management, Rob Hunter, shared with the group that the capital improvements program for the city's water and sewer infrastructure had been moving on time and on budget. However, "it's not about compliance with the federal consent decree," he explained. The city was requesting a four-year rate increase package that would continue funding for the four-billion-dollar overhaul. Hunter explained that the additional investments were yielding substantial savings from reducing leakage and were ensuring the city had a water/sewer infrastructure that would well position it for future growth. ACP members agreed that the capital improvement program must continue if Atlanta is to be a best-in-class city.

June

- Governor Perdue announced the approval of eleven state[135] and federally funded environmental infrastructure projects totaling $91.5 million

- The Atlanta City Council approves the sale of water and waste-water revenue bonds.

July

- The City of Atlanta celebrated National Parks and Recreation Month. During the month, the city recognized the contributions of employees and volunteers who assisted in maintaining public parks and recreation facilities that keep the public parks safe and clean for visitors.

- The Upper Chattahoochee Riverkeeper joined the City of Atlanta to celebrate another major milestone—the completion of sewer rehabilitation work of Sewer Group 1.

  Over 80 percent of the system, almost 1,300 miles of pipelines, has been evaluated, and 300 miles of the system with the most chronic pollution problems have been rehabilitated.

August

- City of Atlanta Office of Cultural Affairs (a division of the Department of Parks and Recreation), and the Atlanta Jazz Festival announced that the 32nd Atlanta Jazz Festival was a winner of the 2009 Greener Festival Award. The Atlanta Jazz Festival's *Go Green* represented an ongoing commitment to sustainability in recycling, transportation, and event administration. The festival had a comprehensive and cohesive

---

135    http://gov.georgia.gov/00/press/detail/0,2668,78006749_78013037_143690275,00.html

waste-management and recycling program and a biodiesel recycling plan for vegetable oil waste from concessions. It encouraged walking, bikes, and public transportation, and it
promoted car pools and ride sharing on the "Clean Atlanta
Campaign" website. The Greener Award was based on a fifty-
six point checklist, which covered green office policies, energy
use and carbon reduction, travel and transportation, support
for green initiatives, waste management, recycling, water use,
environmental protection, and noise pollution.

September

• R.M. Clayton Wastewater Treatment Plant sustained serious
flood damage. The city's largest wastewater treatment facility's primary and secondary treatment capabilities had been
compromised.

• Mayor Franklin and volunteers participated in the inaugural
Summit on Cities[136] of Service in conjunction with National
Day of Service, where large- and small-town mayors encourage Americans to volunteer. She stated that "mayors are on the
front lines and have their finger on the needs of the communities they serve." Events planned for the day included:

  • Cleaning up and beautifying Sphinx Park with Home
  Depot.

  • The City of Atlanta's Office of Recreation hosted thirteen
  facility cleanups with the after-school programs and various neighborhood groups.

---

136    City of Atlanta Online http://174.37.215.145/media/medadv_citiesofservice_091009.
aspx

- MLK Recreation Center kicked off a recycling program with neighboring elementary schools.

- The Bitsy Grant and Sharon E. Lester Tennis Centers and surrounding areas would undergo beautification

- The remaining recreation centers hosted a cleanup of the inside and outside of the facilities.

- The Grove Park Recreation Center was cleaned by the community and a new mailbox was installed.

- The City of Atlanta's Office of Cultural Affairs coordinated efforts to visit multiple sites in three distinct Atlanta neighborhoods, where participants would clean sculptures in the city's public art collection.

- Atlanta's crown jewel (Piedmont Park) was landscaped at the newly restored "Noguchi Playscapes" next to the "Ex-Static" by Atlanta artist Maria Artemis was cleaned;, and the day ended at the urban streetscape artwork "De-code/Re-code Atlanta" at the footsteps of Georgia State University.

- Mayor Franklin, the EPA administrator, Lisa Jackson and other EPA administrators, the Georgia EPD, president and CEO of Atlanta BeltLine Brian M. Leary, and Peggy McCormick, president of the Atlanta Development Authority, announced one million dollars in funding to the City of Atlanta to support the cleanup of sites known as brownfields along the Atlanta BeltLine and other redevelopment corridors. The funds would be used to turn ten local sites contaminated by hazardous chemicals or pollutants into productive properties. This move would help spur redevelopment, secure jobs, and create green space.

October

- Governor Perdue announced the approval of five environmental infrastructure project loans totalling just over $6.7 million.

- Mayor Franklin, council members, residents   took part in the celebration of the the rededication of New Endings in Freedom Park.

November

- City of Atlanta's Department of Parks, Recreation, and Cultural Affairs, District 8 council members, the Trust for Public Land project manager, and PATH Foundation members hosted the Louise G. Howard Park dedication.

- Mayor Franklin  made a quiet exit during the transition for the newly elected next mayor.

- Mayor Franklin left a skeleton team behind to make sure there would be a seamless transition in preparing to assist the incoming mayor. Each department had developed a formal, written transition plan detailing both short-term priorities and long-term strategic issues.

December

- The Department of Parks and Recreation, commissioner Dianne Harnel-Cohen, council members, Georgia Power, the Steering Committee of Morningside, Morningside Place Homeowners Association, and community members celebrated the opening of the Morningside Nature Preserve. The preserve will provide green space, nature trails, and outdoor recreational activities.

- The Department of Public Works, the Atlanta City Council, and community leaders of NPU Unit IV commemorate the beginning of the Metropolitan Parkway Sidewalk Improvement

Project. At an approximate cost of $230,000, the project was scheduled to be completed in  ninety days.

- Mayor Franklin's administration prepared for transition of the mayor-elect.

- Mayor Franklin's cabinet announced their resignations.

Year in Review

Mayor Franklin spent the year focused mainly on operational and administrative work as her administration was winding down. SSES (maintenance, cleaning pipes/repairing broken pipes, assessing remaining work) continued throughout the city from January to December.

In January, Mayor Franklin would send another letter to Congressman Lewis to thank him for his staff's taking the time to meet with her staff and to provide additional details about  the ongoing city projects with supporting details, including economic and environmental benefits. She stated the city had added one hundred thousand new residents since 2000. This growth—a reflection of our success in attracting new businessess and residents—creates environmental challenges related to air quality, water quality, and mobility. The City of Atlanta had a $750 million infrastructure backlog and six billion dollars in future infrastructure needs—needs to be sustainable and generate direct environmental benefits. Mayor Franklin outlined and highlighted upcoming and future "green projects," energy projects, and transportation projects the city would need to get going with financial assistance. She ended the letter stating, "We are convinced that sustainability projects are a critical part of the solution to the economic crisis and will drive a paradigm shift in our economy."

Mayor Franklin hoped that the much needed infrastructure money promised by President Obama would come on time, and it would be the shot in the arm her administration would need. Other mayors who attended the mayors conference at the White House in February included mayors of the following cities: Miami, Florida; Trenton, New Jersey; Akron, Ohio; Charleston, South Carolina; Albuquerque, New Mexico; Tulsa, Oklahoma; Los Angeles, California; Providence, Rhode Island; Denver, Colorado; Southfield, Michigan; Phoenix, Arizona, Washington, DC).

The City of Atlanta released its government report on greenhouse gas emissions, known as the city's "carbon footprint," in partnership with students at Georgia Tech and Valerie Thomas, an associate professor at the H. Milton Stewart School of Industrial and Systems Engineering. She stated, "Having conducted an inventory and commited to reducing emissions makes the City of Atlanta a leader in the state and region and well ahead of federal action on climate change." With the municipal carbon footprint established, the next step would be to develop the Atlanta Climate Action Plan.

In March, transit funds were headed to Georgia. The money was released by the Federal Transit Administration . Georgia representative Hank Johnson (a member of the House Transportation and Infrastructure Committee) stated:[137] "I am pleased just three weeks after I voted for the economic recovery package, we are already beginning to see funds flow directly to communities back home that are suffering under this recession. By funding these crucial infrastructure projects, we will create jobs and help the local economy grow."

Using the ATLStat website as a tool to make the performance measures of operating departments more accessible to the public is another step in the City's efforts to make city government more open and transparent.

---

137    http://hankjohnson.house.gov/2009/03/03-05-transit-funds.ahtml

The tornado that hit the City of Atlanta in May 2008 left the Historic Oakland Cemetery devastated. Thousands of of monuments were damaged or destroyed, hundreds of trees fell and destroyed headstones, and some buildings were also leveled. The city's Office of Parks and Recreation, Park Pride, Historic Oakland Cemetery, and private business partnered and held a re-dedication after a massive yearlong cleanup and restoration culminating in an October road race from the cemetery.

In June, as part of the celebration for National Homeownership Month, the Georgia Housing Summit Alliance hosted a free educational event for consumers. A partnership with HUD, FHA, state and local governments, local housing authorities, nonprofit-organizations, and industry related associations and professionals aimed at promoting and preserving home ownership through ongoing educational assistance and resource tools.

The City of Atlanta celebrated the 145th anniversary of the Battle of Atlanta in July at the Cyclorama Museum with a reenactment of the battle showcasing period-dressed soldiers, weapons, and Civil War stories of the history and heritage of Atlanta.

A record one hundred-year flood had hit unexpectly in September, flooding residences and closing flooded roads, bridges, and other structures. The Atlanta Fire Rescue (the city's fire department)[138] assessed the external damage to public, private, and commercial properties. The assessment was submitted to GEMA (Georgia Emergency Management Association) because of the immediate needs of the residents whose homes were damaged by creeks that flooded in their communities. The city's preliminary numbers showed that 474 homes were impacted by the flooding and approximately 1,900 residents were seeking some level of federal assistance. The Chattahoochee River crested at twelve feet above its normal

---

138    City of Atlanta Online http://174.37.215.145/media/nr_floodassessment_092509.aspx

level after the storms dumped between six and fifteen inches on parts of metropolitan Atlanta.

Mayor Franklin, the battalion chief of the Atlanta Fire Rescue, and the policy advisor on women's issues, Stephanie Davis, announced the launch of a new public safety initiative, SAFE S.P.O.T.S. All City of Atlanta fire stations would be designated a Safe Place off the Streets, offering refuge for victims of commercial sexual exploitation, child abuse, sexual assault, domestic violence, and infant abandonment. The initiative was created by the office of Mayor Franklin in collaboration with the Atlanta Fire Rescue.

Reporting to the federal courts, the Federal EPA and the Georgia EPD ended the year for Mayor Franklin and watershed commissioner Rob Hunter.

Concerning the upcoming mayoral election in November, Mayor Franklin remained mum about whom she would publicly endorse to succeed her.

On the day before the election, Mayor Franklin gave an exclusive interview to CNN host Don Lemon and explained what she was looking for in a candidate. She said that the person would have to have the skills that she and her predecessors possessed. She stated "I will vote tomorrow, and I plan to vote for Kasim Reed."

"Is that an official endorsement?" asked Don Lemon.[139] "That's just telling the truth," Franklin responded. "I'm going to vote for him. I think he has the best set of skills. He has really been there to do some tough things over the state. He has Republican and Democratic support....even though there are other candidates who have obviously some strenghts, I think, through it all, he has the best chance of working in the region and the state."

Lastly, in November, as it related to the mayoral race, it appeared a runoff would be necessary between councilwoman Mary Norwood and former

---

139    http://blogs.ajc.com/political-insider-jim-galloway/2009/11/02/shirley-franklin-says-shell-vote-for-Kasim-Reed-in-ATL-race-for-mayor.htm

state lawmaker Kasim Reed. With 100 percent of precincts reporting after polls closed, neither had the more than 50 percent of the votes needed to win. Norwood had 45 percent while Reed had just 37 percent.

Less than a month before Franklin was set to leave office as Atlanta's mayor, several members of her senior administration team were handing in their resignations[140].

In a memo sent by the commications director, Beverly Isom, the resignations came "as the new administration prepared to transition into their new leadership role."

Announcing their resignations[141] were :

Richard Pennington, Atlanta police chief

Alan Dreher, deputy chief

Benita C. Ransom, commissioner of human resources

Gregory J. Giornell, chief operating officer

Gregory Pridgeon, chief of staff

Jim Glass, hief financial officer

Joe Basista, commissioner of public works

Allison Lehr, City of Atlanta controller

David Edwards, mayor's policy advisor

Lisa Gordon, Enterprise Asset Management director

Bobbie Porche, director of the Office of Constituent Services

Dianne Harnell-Cohen, commissioner of Parks, Recreation and Cultural Affairs

Beverly L. Isom, communications director

---

140    http://www.ajc.com/news/atlanta/franklin-staffers-leaving-paving-226710.html
141    City of Atlanta Online http://174.37.215.145/media/medadv_cabinet_120409.aspx

## Critics

1) After four rounds of layoffs, imposed furloughs on city employees, including police and firefighters, and the closing of several fire stations, Mayor Franklin closed twenty-two recreational centers, shuttered City Hall one day a week, and garbage went uncollected. Critics sneered[142] that the mayor, with all her accomplishments, would leave Atlanta as broke as she had found it. Her remedies, including a property tax increase, had generated bitterness and hostility among constituents.

2) "She's mismanaged this city terribly," said John Sherman, president of the Fulton County Taxpayers Association. "There's been a horrible degree of hype and PR resulting in her pictures on the covers of national magazines. There's much too much hype and not enough results."[143]

3) Would the newly elected mayor, Kasim Reed, continue the aggressive approach to completing the Consent Decree work as mandated?

4) Would Shirley Franklin leave Mayor Reed her "to-do" list on completing what she had implemented, or did she pack up all her belongings, erase all the contacts in her city phone, leave it on the desk, and go out the back door?

## In Summary

The original estimated $3.2 billion needed to overhaul the City of Atlanta's aging infrastructure would surpass four billion dollars. The city used a number of approaches[144] to fund the improvements, including the Municipal

---

142    http://www.nytimes.com/2009/09/08/us/08franklin.html

143    http://seattletimes.nwsource.com/html/nationworld/2009328036_apusatlantamayor.html

144    http://geospatial.blogs.com/geospatial/2011/paying-for-water-and-wastewater-infrastructure:Atlanta-and-Houstin

Option Sales Tax (MOST),[145] the expansion of low-interest, state revolving loans, the sale of local bonds, federal grant funding, tax-exempt commercial paper (commercial loans), current revenue financing, rate increases on water and sewer for residents and businesses, and a $1.2 billion line of credit.

The cost breakdown[146] for the overhaul would be as follows:

One billion dollars to fix combined sewers to improve water quality for discharges from the combined storm-storm and sanitary sewer system, one billion dollars to fix sanitary sewers throughout the system, one billion for additions to water-treatment facilities, and other miscellaneous improvements to the system.

The one-billion-dollar budget for sanitary sewers would set aside $130 million to inspect more than 1,900 miles of sewer lines across the city. That work would be divided into six widely scattered priority groups based on where the city expects to find the worst conditions.

The budget also included $180 million for pipe rehabilitation and $690 million for projects that would add additional capacity. Much of that work won't be set aside until after the inspection is completed.

The remaining one billion dollars would include paying for two treatment plants, $200 million in improvements to existing plants, and $440 million in pipe work.

At the end of her two-term administration in 2009, Mayor Franklin, the woman with the reputation for laser-beam focus, a hardworking technocrat, and with the title "Sewer Mayor" had shown that she had earned the accolades. "Super Shirley" would be another nickname given to her.

145    http://www.atlantaga.gov/index.aspx?page=755

146    *Atlanta Journal-Constitution* archives: D.L. Bennett, June 30,2001 http://docs.newsbank.com/s/InfoWeb/aggdocs/NewsBank/0F47AEA9A66BE868/0D57227E12704560?p_multi=AJBK&s_lang=en-US

In her first two years, the tireless mayor tackled thousands of problems and made remarkable progress on solving most of them.

Mayor Franklin, her administration, and the City of Atlanta had created and established the Department of Watershed Management, which addressed all issues relating to both Consent Decrees.

The first completed Consent Decree separated over 2,000 miles of sanitary and combined-sewer piping that sends storm water to area water-reclamation facilities for treatment then pours it into the Chattahoochee River. By separating the sewers from storm water, and building storage tunnels, waste goes directly to treatment facilities for treatment.

She adhered to the Clean Water Act of 1991 to restore Atlanta's drinking water and protect the rivers and streams with buffer zones and established clean-water standards and training and certification through the Georgia Soil and Water Council Conservation (GSWCC) for all persons working on city contracts through this program.

She had stated, "The future of Atlanta is dependent on clean water. I want a city that is economically viable ten, fifteen, twenty years from now. The sewer is a piece of that. Sewers aren't the sexiest subject matter, still, their importance can't be overlooked".

Mayor Franklin, in her effort to make the City of Atlanta more green, went from having one of the lowest percentage of LEED-certified buildings to one of the highest.

Work continues to date throughout the city under Mayor Reed as new contracts are signed for the various phases of work in upsizing sewer line piping in outfall areas to alleviate overloading, replacing aging manholes, water line installation, and water meter replacement.

Mayor Franklin was faced with a 2014 mandate that could no longer be ignored by the EPA's Clean Water Act and the "empty can" could no longer be kicked down the street. It was predicted that by 2014, at the program's end, Atlanta could be holding close to four billion dollars in debt that would take forty years to pay off. The final payback, with interest included, would be more than seven billion dollars, officials say.

She vowed that the Consent Decrees would be completed by 2012, two years ahead of schedule.

Mayor Franklin took the task of overhauling the city's infrastructure at hand and attempted to completely overhaul every aspect of it to the best of her abilities, and along the way started partnerships with downstream counties. The accomplishments that followed during her tenure and after her departure have defined her administration.

She had the confidence to know that her track record and legacy would be judged by her accomplishments. She had the courage to make the "tough choices" to ensure the city's financial well-being. She stated, "If you have to worry about your legacy, you don't have one."

Mayor Franklin had left her mark on the hearts and minds of the citizens and business communities she served in Atlanta, and she has left a strong, lasting legacy, maybe never to be duplicated.

"I hope that we have set an example," she stated in an interview with Lynne Anservitz[147] of Business Library. "We've made significant progress. We've drastically reduced sanitary and combined sewer overflows across the city. We've separated the sewers into three sewer basins, leaving only the downtown area with combined sewers. We've built more than 120 miles of new water mains. We've inspected more than 1,000 miles of sewers and rehabbed about a quarter of them. Because of this effort, the Chattahoochee River is cleaner than it was 10 years ago. I hope we have set an example that future administrations will follow. I think people understand for the most

---

147    http://findarticles.com/p/articles/mi_m0KXG/is_2_10/ai_n27953129

part what we're trying to do here. They know that sewers are an intrinsic part of Atlanta's quality of life. They realize the work had to be done". "I was a one-person chant, a drumbeat for infrastructure".

She stated,[148] " I hope that I've made people proud by my honesty and hardwork, and being truthful and forthright. And that I came to work every single day, worked hard to address the challenges that face the city, from infrastructure problems to crime. So my pledge was to do my best every single day so that this would be a great place for every one to live."

"I had no choice," she admitted. "Events and circumstances dictated my agenda."

During Mayor Franklin's tenure, Atlanta had run five balanced budgets in a row, including a surplus of seventeen million dollars in 2005.

She passed the toughest ethics law in Georgia.

"I believe you recruit the best talent you can, and go with answers that are databased and researched and targeted to best practices," she said. "If you can sell the best practice, you can sell anything".

Mayor Franklin would state, "There are people, black and white, who have said to me, 'I don't think a woman is tough enough to do this job.' It was uncommon for business people and civic leaders to say, 'So far, you're do-ing well.' Now, isn't that an interesting comment. Remember, I studied sociology and psychology, so I'm paying attention to the interaction as much as the words."

"The business community had always been willing to contribute formally or informally," said Art McClung, director of Atlanta operations for Georgia Power. "Mayor Franklin has created an environment of opportunity for partnerships between the city, the business community and citizens. It's a win-win situation for all players." He added, "Real leaders aren't willing to

---

148    http://www.tbsstoryline.com/conversations.html

wait forever to make something happen. Shirley Franklin isn't waiting and, as she's demonstrated, she's more than willing."

Public-private partnership was a hallmark (bread/butter) of her approach to governance. She would state, "I learned early on, particularly from Andy Young, that the business community is an important part of what makes a city great. These are the people who invest and reinvest."

"She's able to operate both strategically and tactically," says Peter Aman, Mayor Franklin's pro bono consultant. "We work with a lot of CEOs. She's as good as any fortune 100 CEO I've ever worked with."

She became the bearer of bad news no one wanted to hear, and at one time stated, "I have a bull's-eye on my chest. I knew that when I ran for office." But over time, that would fade and before she left office even the city council made amends and let her know that they should have listened to her years ago.

Mayor Franklin admitted that "the job can be grueling and tiring and gruesome, and not everyone has the sheer determination to plow through the mess. I do. I have the sheer will."

Sally Bethea of Upper Chattahoochee Riverkeeper commented that "without Mayor Shirley Franklin's support and encouragement, the Clean Water Atlanta program never would have happened." While continued investment must be made to finish all the work by 2014, the city and its neighborhoods are already benefiting, thanks to a healtier environment.

Was Shirley Franklin an effective mayor?[149] She refused to speak with reporters about her legacy. "I will not speak about my legacy," she was quoted saying on her way out of office. The article goes on to suggest she focused or reflected maybe more on the negative things and disappointments that

149    http://www.associatedcontent.com/article/2604557/was_shirley_franklin_an_effective_mayor_for_atlant.html Associated Content from Yahoo- Rhonda Turner, Jan 20, 2010

happened during her later years, as in 2008 or 2009, when council members at a press conference called by the mayor in the atrium of City Hall wanted to let the citizens know publicly that they wanted to distance themselves from her proposed budget while her staff looked on. "Franklin is catching flak from nearly every side these days[150]: from the council, where even longtime allies are voicing disappointment with the mayor's oversight of city finances. From the public—or at least those who show up to open-mic hearings and post angry comments on newspaper websites. And even from the boosterish daily newspaper. The article stated that when Cynthia Tucker of the *Atlanta Journal-Constitution* wrote a column headlined "Fiscal meltdown a blot on Franklin's tough tenure" the event elicited a collective gasp from local media watchers. Surely a corner had been turned.

All political careers have their ups and downs, but Franklin's downs have mostly come in the past couple of years, some say with the shooting of Kathyrn Johnson and government-related scandals, while others cite political missteps and a declining real estate market. Still others point to the much ridiculed Brand Atlanta PR campaign, and lastly her own personal family issues.

For me, most of Mayor Franklin's critics were doing nothing more than hitting her with "cheap shots," hitting below the belt, and wanting to poke their fingers in her eye. Bloggers would selfishly unleash their anger back and forth on every issue the city would face—all without merit.

In another article[151] entitled "Atlanta's mayor faces rocky road out of office," the writer states: "It's not easy for any big-city mayor during a recession. Still, for Franklin the fall from grace has been particularly steep. The Democrat is no longer routinely mentioned as a candidate for statewide office. With roughly five months left until a successor is chosen, Franklin

150    http://clatl.com/atlanta/the-chinks-in-shirleys-armor/Content?oid=1273616

151    http://seattletimes.nwsource.com/html/nationworld/2009328036_apusatlantamayor.html

said she's not spending much time second-guessing the tough decisions she's had to make or fretting about her legacy. "I find being depressed and regretful are time-consuming and unhappy and contentious," she said in a recent interview with the Associate Press. "As a working, single mother for most of my life, I've just decided there are some things that don't have a place in my everyday life and one of them is crying over spilled milk".

Mayor Franklin's luster in her mind was starting to fade, and she was not as positive about things as she was when she came into office like a newly charged locomotive, the great reformer of Atlanta's city government. She should have not taken her critics for nothing more than clanking empty cans.

Mayor Franklin always wore a flower on her suit jacket. Why the flower? She stated in an interview with Kate Carter[152] that she loved flowers and even had a small garden at one time, but the flower to her symbolized femininity.

Magazines, newspapers, and other media outlets loved featuring articles about her, including the following: the *New York Times*, *Ebony*, *Jet*, *Fortune*, *Forbes*, *BusinessWeek*, *Seventeen*, The Associate Press, *Essence*, Savoy, Terry College, Silke Endress, the *Wall Street Jounal*, *US News & World Report*, *Atlanta Business Chronicles*, *Black Enterprise*, The Huffington Post (an Internet newspaper), BlackPast.org (an Internet publication), Politifact.com, CNN Tech, US Mayor newspaper, *Governing* magazine, *Zimbio* magazine, *Bluprint* magazine, TBS Story line, *Georgia Trend* magazine, *Invest Atlanta*, Iowa State University (archive of Women's Political Communications), the *Christian Science Monitor*, The *Seattle Times*, *Creative Loafing* magazine, *Waterkeeper* magazine, and the *Atlanta* magazine. Even the *Atlanta Journal-Constitution* (the local newspaper) cheered her efforts in their articles.

---

152    http://www.divinecaroline.com/22365/27266-mayor-shirley-franklin-straight-talking-lady#ixzz1fsg9Q1XQ

## Her accomplishments are vast:

She was elected incoming chair of the Women Mayors Caucus of the United States Conference of Mayors in 2002, and awarded an honorary Doctor of Laws from her alma mater, Howard University in 2002.

In 2003, Mayor Franklin hosted Atlanta's first Green Fair in 2003 with Energy Star and Home Depot at City Hall.

In 2004, she was named Public Official of the Year. She launched "Welcome to Atlanta- NASCAR Weekend"; she was named one of ten Grand Marshalls for the 2004 MLK Jr. Holiday March and Rally.

In 2005, she was named Top Mayor of the Year; named among World Mayor Top Ten of 2005; won the John F. Kennedy Profile in Courage Award; named among Five Best Big City Mayors in America; named American City & County Municipal Leader of the Year; was named One of *Esquire* magazine's Best and Brightest. Atlanta was named a Top "Mid-Sized City" Arts Destination, and the city ranked number two among Top 10 Tech Cities.

In 2006, she was named one of 100 Most Influential Atlantans, and Americas Best Leaders; she received the Southern Institute for Business and Professional Ethics Advocate Award.

In 2007, she was named Most Respected leader; was awarded an honorary Doctor of Law from the University of Pennsylvania. She was inducted into the International Civil Rights Walk of Fame. *Newsweek* magazine named her one of the women to watch for in its Women & Power issue.

In 2008, she served as coChair of the Democratic National Convention, where Barack Obama was nominated for president of the United States. The White House Project named her one of its "8 in 08" likely to run/or be elected president. She attended the 6th Annual Warrick Dunn Gala in 2008 to celebrate helping moms with affordable single-family homes.

In 2009, she won the Four Pilar Award (Tribute). She was appointed to an ad hoc Department of Homeland Security special task force; she received the University of Pennsylvania IUR Urban Leadership Forum Award; was named by secretary Janet Napolitano to the Homeland Security Advisory System Task Force; and she was named One of the Nation's Top 'Eco-Mayors' of 2009.

In 2010, she was named William and Camille Cosby Endowed Professor in the Social Sciences 2010 (Spelman College). She also holds a professorship at Spelman College. She received the NAACP leadership award, the YWCA Woman of the Year and Woman of Achievement awards.

These accolades don't include her other city and humanitarin accomplishments to fight homelessness by building a center for the homeless,[153] She held global conferences on economic development, helped the formation of the Civil and Human Rights Center partnership effort, was instrumental in HUD's awarding more than $9.4 million in grants to supprt dozens of homeless programs in Atlanta, DeKalb, and Fulton counties, partnered with AGL Resources and Atlanta Gas Light to donate 500 blankets to the Atlanta homeless community, partnered with Popeye's Chicken to donate turkeys to Atlantans in need, and celebrated annual Fathers Day "real men cook for charity." She addressed quality-of-life issues for Atlanta residents, she held global gender equality conferences, she helped with the coordination and  logistics with the Red Cross and community leaders to provide assistance to all relocated evacuees of Hurricane Katrina with a disaster assistance program, and initiated the Old Navy's "Field Trip 4 Fun" shopping spree for young hurricane victims. The City of Atlanta hosted the Annual Children's Holiday Festival. She named a new police chief, had a new City Hall and municipal court building built, added more than 300 sworn police personnel during her administration, created the

---

153    Center for Civil & Human Rights (Atlanta, Ga.)

Weed and Seed program, reducing the city's crime ranking, held an annual CitySafe Summit to address illegal drugs and guns in the city, and cleaned up corruption in city services and established a code of ethics for city workers. She held foreclosure-prevention programs, helped reopen Atlanta City Employees Credit Union and celebrated the rededication of the Neighborhood Charter School in Grant Park. She initiated numerous city job fairs, attended the annual mayor's youth program "Tee Up for College" golf outing, initiated DREAM JAMBOREE, the Southeast's largest college and career fair, Youthfest 2006, and summer internship for students. She initiatied the Mayor's Youth Program (MYP), preparing teens for the health-care industry while still in high school, welcomed the birth of VIBE Musicfest, and initiated job training and the Atlanta Workforce Development Agency that included a million-dollar shootout golf tournament. She launched the first City-Wide Reading initiative. She held a 2004 voter registration drive in the City Hall atrium, announced a Voting Rights Act march with the Rev. Jesse Jackson, and expedited the permits policy for construction of one-and two-family homes for low-income housing. She is instrumental in teaching and lecturing young women at Terry College and Spelman College, She kicked off the Statewide Child Safety Project with the Conference of Black Mayors. She jumpstarted the expansion of the Hartsfield-Jackson airport's fifth runway (a $5.4 billion expansion), with three new checked baggage security screening systems, initiated negiotations for Delta Airlines Nonstop flights to China. She started the broadband initiative ( wi-fi network) for the city and airport, was instrumental in the development of the BeltLine, and developed public-private marketing business partnerships for the city to position Atlanta as a key tourism and business destination. She declared Atlanta first "City of One" to fight extreme poverty and preventable diseases. She secured partnerships with local companies to distribute fans for the elderly (stay cool campaign) in the summers. She secured one million dollars  for a toilet-rebate program. She held an annual Mayor's Cup Golf Tournament to raise money for Camp Best Friends and Mayor's Bowl: Strike for Camp Best. She assisted with the cleanup of Historic Oakland Cemetery in 2008. She held vested interest in the Fort McPherson Redevelopment for the city, and started a New Sister City Partnership between Atlanta and Fukuoka, Japan. She started an annual IT Expo, restored the faith with the Upper Chattahoochee Riverkeeper, added 1,179 acres of greenspace since 2001 (a 37 percent increase), boosted the development of the Peachtree Corridor, created the Brand Atlanta launch campaign to energize economic growth, and helped save Grady Memorial Hospital from closure. She appointed city judges, secured the papers of Dr. Martin Luther King Jr. (keeping them in Atlanta), balanced the city budget three times in office, and improved the city's bond from negative to stable. She got approval for a 1 percent sales-tax hike and a 50 percent bump to property taxes. She consolidated the city and county 911 services, and she even cut her own salary by forty thousand dollars all in her effort to streamline city government.

Her involvement with the Atlanta Public Schools included "Next Step,"[154] which she created for every Atlanta-area high school senior to get engaged in planning for his or her future beyond high school. She helped the Atlanta firefighters become "Best in Class" in using safety technology. She had a new state-of-the-art judicial case-management system (court overview) incorporated into the new municipal court building. Mayor Franklin was also chosen to be featured in a new series of biographies for children "Meet Shirley Franklin, Mayor of Atlanta," the first release in a series being published by Gallapade International of Peachtree City.

Mayor Franklin wrote an article in *Invest Atlanta*[155] about the most recent accomplishments of Hartsfield-Jackson airport:

---

154    http://www.mayorsyouthprogram.org/pdfdoc/MYP%20Brochure.pdf

155    http://www.investatlanta.com/atlCompAdvan/transportationHub.jsp

- We constructed more than 1 million SF of cargo ramp space, which will accommodate 9 Boeing 747-400s simultaneously,

- We added 375,000 SF of modern air cargo handling space,

- We now have 80 flights per week to Latin America, and Atlanta wants to be the home for the Sectretariat of the Free Trade Agreement of the Americas, and

- We have 32 flights per week to Asia; in fact, 22 of these flights are providing wide-body, all cargo service. Air carriers such as China Airlines, Polar Cargo, and Japan Airlines are providing service and jobs to the local economy.

- Today, we have almost 40 international all-cargo flights per week.

Mayor Franklin was big on volunteering too.

She serves on the Georgia Municipal Association. She partnered with Home Depot and Habitat for Humanity in a program where volunteers built "Community of Dreams," and "United we Serve" to help raise one million dollars to fund housing and park projects. She had volunteers for the 2006 homeless census count. She helped the city receive federal grants for seat-belt safety. She got high school students to buy into cleaning up their neighborhoods to earn credit, and built a community garden that taught caring for the environment. She initiated City workers' volunteering to clean up parks and along the Chattahoochee River. She participated in the Life/Walk for Prostate Cancer. She initiated volunteering to "adopt a ramp" in the city where organizations kept an on/off exit ramp clean on the weekends. She held the annual Earth Day (an environmental initiative for a greener city), a citywide community-based project. She joined forces with civic and environmental groups to volunteer their time and service to help the City in her causes, and travelled with CARE to South Africa. She participated in the inaugural Summit on Cities of Service (September 2009) to join with other large and small-city mayors to encourage Americans to volunteer.

She cochairs the Regional Commission on Homelessness, serves as vice-chair of the Center for Civil Rights and serves on the board of the United Nations Institute for Training and Research.

She sits or has served on the boards[156] as president and 2nd vice president of the Georgia Municipal Association (GMA); chair of the Atlanta Development Authority, and chair of the Brand Atlanta Campaign; on the Board of Trustees of the United Nations Institute for Training and Research; chair of the Board of Trustees of CIFAL-Atlanta, Delta Air Lines, and Mueller Water Products Inc.; member of the National Conference of Black Mayor's Business Council; secretary of the Board of the Atlanta Regional Commission; chair of the Ethics Committee and Secretary of the Executive Committee for the City of Atlanta, the United Way of Atlanta, and was appointed to the Georgia Regional Transit Authority. She previously served as the founding vice president of the Georgia Regional Transportation Authority (GRETA). She was a member of the Mayors Against Illegal Guns Coalition. She joined the Atlanta Committee for the 1996 Olympic Games Inc (ACOG) as the top ranking female executive, a senior vice president for external relations,[157] a primary liaison with labor unions, civil rights groups, neighborhood organizations, and environmentalists. She became a majority partner in Urban Environmental Solutions in 1998. She even owned her own management and consulting firm (Shirley Clarke Franklin & Associates) for public affairs, community affairs, and strategic planning, where she was instrumental in strenghtening ties with the local community in the East Lake Redevelopment and golf course. She

---

156    The Biography of Mayor Shirley Franklin

157    http://www.uimonline.com/index/webapp-stories-action?id=19&archive=yes&Issue=2005-atlanta-mayor-shirley-franklin

served on the boards[158] of several civil and cultural organizations, including the Atlanta Symphony Orchestra, the National Endowment for the Arts, Spelman College, and the National Urban Coalition.

Not bad for a woman who stands five feet and one inch tall who earned a bachelor of arts degree and master's degree in sociology.

In 2010, after her departure as mayor, she became an advocate for broadband access for the underserved. She was named by the Alliance for Digital Equality as a special senior policy adviser to find and promote private-public partnerships to bring high-speed Internet services to underserved communities.

Three former mayors of major metropolitan cities were encouraging the Federal Communications Commission (FCC) in 2011 to approve AT&T's proposed purchase of T-Mobile USA.[159] "Technology and mobile broadband are essential to the development, sustainability, and the future of our economy," said Ronald Dellums of Oakland, California, Shirley Franklin of Atlanta, Georgia, and Douglas Palmer of Trenton, New Jersey. "This merger takes us one step closer to ensuring that African Americans have greater access to the transformative power that wireless technology offers and the opportunities that come with being connected to an ever-growing digital society."

Mayor Franklin now works as CEO of Purpose Built Communities, returning to her business roots.

---

158    http://www/lasentinel.net/ShirleyFranklin

159    http//politics365.com/2011/06/24/former-mayors-say-attt-mobile-merger-extends-transformative-power.htm

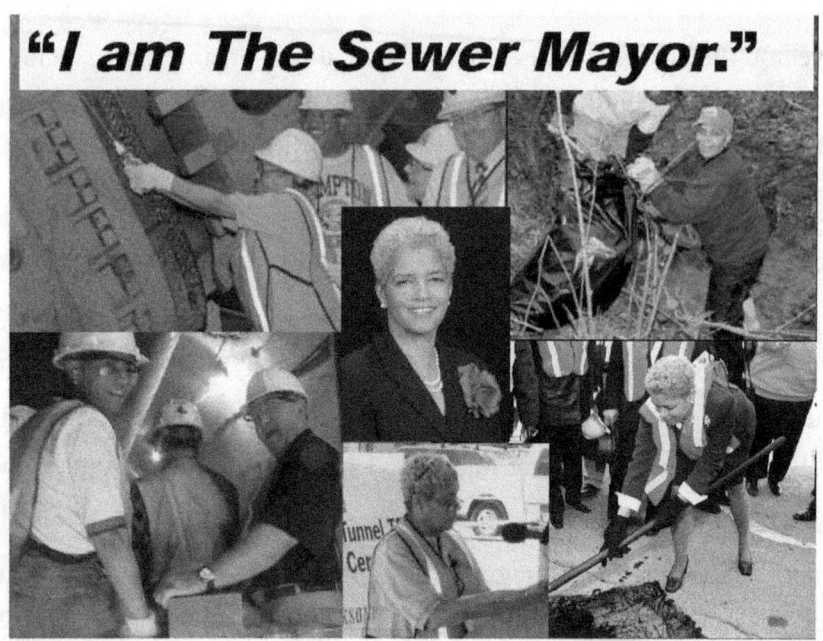

http://www.pueblo.us/DocumentView.aspx?DID=480

# Part 4 - Department of Watershed Protection

The Bureau of Watershed Protection has responsibility for environmental compliance programs that include green-space protection, stream-bank stabilization, industrial pretreatment, watershed monitoring, interjurisdictional flow monitoring, grease management, erosion and sedimentation control, storm-water compliance, and land development regulations.[160]

## Flow Monitoring

Flow monitoring is a vital component of assessing amounts of flow traveling in sanitary sewer pipes, manholes, creeks, and outfalls. Sampling is a large part of the program. Strategically placed flow meters of all different sizes are placed throughout the sewer system, tunnels, and creeks around the city, the sanitary sewer treatment and water-Treatment facilities.

The Department of Watershed Management (DWM) and Clean Water Atlanta (CWA), with the Project Management Team (PMT) consultants[161] evaluate the City of Atlanta's wastewater collection and transmission system. The system comprises ten sewer basins, including areas of 1,600 miles of separate sanitary sewer system and approximately 400 miles of combined sewer system, The DWM, CWA, and the PMT are charged with managing and building a hydraulic model using Infoworks CS software for the city's ten sewer basins to quantify the amount of inflow and infiltration

---

160    http://www.atlantawatershed.org/bureaus/storm/WPhomepage_a.htm

161    Alberto Bechara, PE.

(I/I),[162] and determine system defects and hydraulic performance under the dry-weather and wet-weather flow conditions for the system. This is achieved by implementing a network of permanent and temporary flow meters and rain gauges and model calibration of several storms, matched with measured-flow meter data. Assess the performance of the sewer system using a 2 year -3 hour design storm as well as develop alternative model scenerios to cost effectively, address capacity limited issues and I/I (inflow and infiltration) reduction component and rehabilitation of the system and help generate an SSO remedial report for the EPA and the Georgia EPD showing various alternative solutions such as upsizing the current system, storage above ground, and underground tunnels to eliminate surcharge and overflow conditions.

Flow monitoring in creeks is critical to identifying rising creeks' flow velocity and possible flooding. Remote monitoring in creeks requires sensors that detect and record flow rate, daily activity, etc.

Flow monitoring in sewers is critical to logging capacity in the pipes and in the manholes, and at wastewater-treatment facilities.

How storm-water runoff affects you?[163]

When water from rain and melting snow runs off roads into our rivers, it picks up toxic chemicals, dirt, trash, and disease-carrying organisms. Studies show that this storm-water pollution rivals sewage plants and large factories as a source of damaging pollutants in our drinking water and at our beaches.

---

162    Inflow: water entering the system from above the manhole surface (i.e., rainwater, puddling, submerged manholes, etc.); infiltration: leakage or egress of sewage or water flow into pipes from the surrounding area.

163    http://www.nrdc.org/water/pollution/storm/chap7.asp

The goal of storm-water monitoring is to capture runoff during a storm event.[164] Storm-water monitoring systems can vary depending on compliance regulations, research criteria, and site deployment challenges. The City of Atlanta works with the National Oceanic and Atmospheric Administration (NOAA) Storm Tracker via satellite.

## Remote Monitoring

The City of Atlanta's Office of Watershed Protection entered into contractual agreements with six surrounding jurisdictions that are tributaries to Atlanta's owned and operated treatment facilities. There are approximately 120 different jurisdictional sewer crossings around the Atlanta area and Atlanta has elected to measure the flow at the various locations. One of the systems the city has elected to use is cellular and solar technology.

Cellular Wireless Water-Flow Monitoring Program[165]

Multi-Tech was the company chosen to establish a string of AT commands that enable software (Flowlink) to communicate with various types of open channel meters (ISCO) in the field. The reason is that the city would not need an actual person on-site in the field, eliminating the expense of hiring and training extra personnel.

Remote monitoring in sewers requires installation of a sensor inside the diameter of the pipe near the manhole. This sensor sends information to a remote computer installed outside the manhole either on a tree or a pole high enough above ground to send a satellite signal of the data back to the command center. Another separate sensor is installed inside the manhole and connected to the sensor in the pipe. This sensor is located high enough inside the manhole to detect when flow inside the manhole has risen to the

---

164   http://www.ysi.com/media/pdfs/E100-Stormwater-brochure.pdf

165   www.multitech.com

height of the sensor and starts to record its flow. This monitoring is vital in determining when flow in the manhole is rising or may surge.

The Remote Monitoring Division is responsible for generating data used for billing surrounding municipalities that use Atlanta's facilities to treat their wastewater.

## Erosion and Sediment Control

The City of Atlanta as a component of the Consent Decree was mandated to protect its streams, and stream banks. The natural vegetation along creeks provides cover to the ecosystem within the state waters and must be protected. Vegetation holds soils in place and slows the flow of water, reducing erosion and trapping sediment.

Erosion[166] from construction activities is of concern for the state and for contractor crews working near creeks doing aerial sewer crossings perpendicular to the creek, point repairs, open-cut excavation, pipe bursting, or manhole replacements. The sediment[167] that comes from disturbing the soil has to be contained at the creek by providing protection to the creek with stream buffers.

The state contends that within thirty days after the date of the Notice to Proceed, the contractor shall submit a narrative description, working drawings, and schedules for proposed controls for temporary erosion and sedimentation to the local authority and engineer for approval. The description and working drawings shall meet the requirements of the Georgia Erosion and Sedimentation Act of 1974 (as amended) and local soil and sedimentation control ordinances.

---

166    The process by which the land surface is worn away by the action of water, wind, ice and gravity

167    The process by which the eroded material is transported and deposited by water, wind, ice, and gravity

All erosion and sediment control devices, including check dams,[168] shall be inspected by the contractor or local authority agent on a weekly basis and after each rainfall occurrence.

One method most commonly used by contractors near creeks and streams, and a requirement of the state, is a double row of silt fence (Type C) with wire backing and metal posts with a minimum two feet of straw (staked bales) placed between the fencing to control erosion and sediment. This fencing should be checked and maintained regularly once installed due to the movement of soils eroding during heavy rains and when creeks rise.

What is a State Water[169]?

According to the Georgia Erosion and Sediment Control Act of 1975, "State Waters" include any and all rivers, streams (wet or dry), creeks (wet or dry), branches, lakes, reservoirs, ponds, drainage systems, springs, wells, and other bodies of surface or subsurface water, natural or artificial, lying within or forming a part of the boundaries of the state that are not entirely confined and retained completely upon the property of a single individual, partnership, or corporation.

In Georgia, the Georgia Soil and Water Conservation Commission (GSWCC) is the governing body. They ensure compliance, permits, training, best management practices (BMP), and penalties. Contractor crews working on the Clean Water Atlanta project are required to obtain erosion and sediment control traing of minimum Level 1A (Certified Personnel) or Level 1B "Advanced" Fundamentals within one year of starting work.

When you are looking at drawings that have designed erosion and sedimentation BMPs for a site, look for the stamp that says a certified design professional Level II has certified the drawings.

---

168    A small dam, either temporary or permanent, that reduces erosion and gullying and allows sediments and pollutants to settle

169    www.erosiontraining.com (stream buffer requirements)

## What is the NPDES?

The National Pollutant Discharge Elimination System from the Clean Water Act (CWA), formerly called the Federal Water Pollution Control Amendment of 1972 under the EPA, has implemented pollution programs such as setting wastewater standards for industry and water-quality standards for all contaminants in surface waters.

The state of Georgia's Department of Natural Resources, Environmental Protection Division, Permit No. GAR 100002 was established for infrastructure construction projects. This permit authorizes, subject to the conditions of this permit, the following:

1. All discharges of storm water associated with "infrastructure construction projects"[170] that will result in land disturbances equal to or greater than one acre.

## Why is the NPDES Permit Important?

The NPDES permit program controls discharges. **Point sources** are discrete conveyances, such as pipes or man-made ditches. Industrial, municipal, and other facilities releasing discharges must obtain permits if their discharge goes directly to surface waters.[171]

Municipal wastewater-treatment plants are amoung the most significant point sources regulated under the NPDES program in the Chattahoochee River basin, accounting for greater than 96 percent of the total point-source effluent flow (exclusive of cooling water). These plants collect, treat, and release wastewater.

---

170   Construction activities that are not part of a common development, and could include the installation of roadways, conduits, pipes, pipelines, substations, cables, wires, trenches, vaults, manholes and similar or related structures

171   www.epa.gov

Other sources referred to as **nonpoint sources** consist of mud, litter, bacteria, pesticides, fertilizers, metals, oils, grease, and a variety of other pollutants washed from rural and urban lands by storm water. The most significant nonpoint sources are those associated with precipitation, washoff, and erosion, which may move pollutants from land surface to water bodies. Both rural and urban land uses can contribute significant amounts of nonpoint pollution.

The EPD's NPDES permit program provides a basis for regulating municipal and industrial waste discharges, monitoring compliance with limitations, and approriate enforcement action for violations.[172]

If you plan to conduct activities that may impact the state's natural resources and environment, then you may need one or more environmental permits.

Sediment is a water pollutant. Storm-water runoff is the number one cause of impaired water quality in Georgia's streams, rivers, and lakes. Storm-water permits apply to all sites or common developments which disturb one acre or greater, including smaller tracts within a common development that is larger than one acre where storm water may leave the site.

Types of NPDES Permits include:

Discharge to surface water: a site-specific NPDES Permit is required for discharge of wastewater from a facility to surface waters of the state.

Nonprocess wastewater: includes cooling water, reverse osmosis or softener-rejected water, boiler blowdown water, or cooling tower blowdown water.

Process wastewater: includes any water that comes in contact with the product, coproducts that can become a wastewater or Clean In Place (CIP), and other wash and cleaning water.

---

172    http://www.gaepd.org/files_PDF/plans/chatt/chatt-4.pdf

Treated sanitary wastewater: is subjected to secondary treatment standards found in Title 119, Chapter 21. Form 1 and 2A can be used for faciliites discharging only treated sanitary wastewater.

## Geographical Information System (GIS)[173]

The City of Atlanta had contracted the Program Management Team (PMT) under the Clean Water Atlanta program to assist them with compliance to a Consent Decree issued by the US Department of Justice in 1999.

Working side by side with the Department of Watershed Management, Engineering Information Division, Khafra Engineering was tasked with building an enterprisewide GIS system. In an effort to improve the city's mapping system, CADD-based sewer drawings were converted to a GIS format, thereby transforming a purely graphical representation of the city's collection system to a multi-dimensional information-based system. These data are stored in an Oracle database and accessed through ESRI's Spatial Database Engine (SDE). Enterprisewide GIS is a term used to indicate that many different users can access and update a GIS database simultaneously without loss of information or corruption of datasets. Khafra has also implemented a system whereby this GIS information could be accessed/distributed via customized Internet mapping interfaces and web-based tools.

The GIS department[174] is tasked with collecting information on maps and data in the city of Atlanta, including:

1) Lot boundary maps

2) Annexation

3) Interactive maps

---

173    Abstract: Municipal Asset Management Using GIS and Internet Mapping: The Atlanta Story by Clare Brown  and Keith Toomer

174    http://gis.atlantaga.gov

4)  Zoning maps

5)  Property information search

6)  Document archive

7)  GIS data catalog

8)  Map collection

9)  Department of Watershed Management GIS

*Notes:* GIS mapping is instrumental for identifying coordinates on manholes and storm structures, locating pipeline segments on streets, identifying fire-hydrant locations, or structures.

## Grease-Management Program[175]

As part of, and in accordance with the First Amended Consent Decree, the City of Atlanta is responsible for the management, operation, and maintenance of its sanitary and sewer systems. The grease-management program was established for permitting, maintaining, and monitoring requirements for the discharge of fats, oil, and grease (FOG) from food-service facilities into the city's wastewater-collection system.

A City of Atlanta ordinance was passed in September 2003 because of the city's experience with sewer overflows as a result of blockages caused by the discharge of fats, oils, and grease by food-service, sales, and processing businesses throughout the Atlanta's sewer system.

The city adopted Ordinance No. 2001-5 and revised the Code of Ordinances, Part 154, Division IV, Section 154–297 to protect its sewer system from the damaging effects of the discharge of fats, oils, and grease. All fees collected in accordance with these ordinances would be deposited into a fund.

---

175    http://www.atlantawatershed.org/bureaus/storm/WP_Grease%20Management_a.htm

All food-service facilities are required to submit a complete permit application, pay permit fees, and be available for inspections.

The City of Atlanta defines a facility to mean a building or a location where food service, sales, or processing occurs, as identified by a number for public streets used as directions for delivery. This also includes wastewater discharge from restaurants, schools, nursing homes, day-care centers, and other public facilites.

Fats, oils, and grease are a consequence of cooking from animal fats and vegetable oils found in meat fats, lard, butter and margarine, cooking oil, shortening, food scraps, and baking goods. This grease is frequently poured down the sink drain. The warm oils are in liquid form when poured into the drain pipes, but as the liquid cools, the grease solidifies and causes buildup inside the pipes and becomes a hardened mass. This buildup inside the pipe restricts the flow of sewage and clogs the pipes.

The City of Atlanta's grease-management program maintains that all food-service, sales, and processing establishments discharging fats, oils, and grease to the city's sewer collection system and water-pollution control facilities shall operate and maintain a sufficiently sized grease trap necessary to achieve and maintain compliance.

The grease-management program would address:
- Grease compliance information
- Permits and grease-trap requirements
- Plan review/handout
- Grease haulers
- FOG best management practices
- Residential grease

- Food-service wastewater discharge

- COA standard grease-trap design handout

Note: The grease-management inspection has kept over 1.5 million gallons of grease out of the system since the first quarter of 2009.

# Part 5 - Department of Engineering Services

The Bureau of Engineering Services (BES) is responsible for managing the DWM's capital improvement program, including designing and constructing projects to comply with the city's Consent Decrees, administrative orders, and other improvements to the city's water and sewer systems. The Bureau of Engineering Services provides design, consulting, and project-management services and is responsible for controlling construction costs and quality.

Under the Department of Watershed Management's Program Management Team (PMT) as part of the overall assessment of the sewer system the Sewer System Evaluation Survey (SSES) program came about to address the current conditions of the system as a whole.

## 1. The SSES Program-

The City of Atlanta conducted a thorough, comprehensive evaluation of the sanitary sewer system.[176] This is the first step in the city's massive sewer-system rehabilitation, repair, and enhancement program. Work started in July 2002 on mapping the network of sanitary sewer lines in the city.

The Sewer System Evaluation Survey (SSES) program would involve the inspection and repair of 2,200 miles of sanitary sewers throughout the city that were cracking, leaking, separated, or otherwise damaged. These pipes

---

176    http://www.cleanwateratlanta.org/SSES

address the entire system in the City of Atlanta, including assessing manhole conditions. Throughout the city, there are a variety of different types of pipe materials connected together over the years (vitrified clay, concrete, cast iron, and ductile iron).

The program includes hiring competent and qualified contractors, working different sections of the city where engineers expect to find the worst pipes to perform a variety of services, and using various repair techniques to detect and fix leaks in pipes. The techniques include smoke bombs and dye tests to find cracks and leaks as well as infiltration in the pipe from roots, separation, sagging, and upheaving. The objective is to repair damaged service connections to residential homes and businesses and to inspect the entire sanitary sewer system using closed circuit television (CCTV) video inspection. The inspection information is sent to city engineers, who review and assesst the data to determine if sewer lines will need upsizing (larger diameter pipes) and assess if manholes should be repaired, replaced, lined, or scheduled for improvements in the future.

Sewer sheds were catorgized into six sewer groups. The designation SG-1 means those sewers are believed to be the most critical in needing repair or are the most problematic. SG-6 means those sewers are the least critical in needing repair in conjunction with other major projects.

Once engineers inspect the CCTV video of a particular line segment (manhole to manhole), then they discuss which repair method would best be used to repair or replace the damaged pipe given the condition of the pipe and the location of pipe (i.e., near a creek, in the outfall, under a bridge, near tall buildings, conjested downtown areas, or on residential property).

The next step would be to write up work orders for the construction managers to give to the contractors to perform the work.

## 2.   Maintenance

Maintenance is the core aspect of fixing the infrastructure. Various techniques are used in the inspection, repair, and replacement of damaged pipes, and in assessing manholes and storm structures. Defects in pipes and structures that lead to further decay have to be addressed. There are two types of defects:

Point- defects occur at a precise point in the sewer

Continuous- defect runs along the sewer without interruption

Inspection is a very important part of maintenance. Pipes and manholes can be visually inspected (if exposed). A CCTV camera can be used, or sonar-inspected for large-diameter pipes that are full with flow.

CCTV- Closed Circuit Televised (for pipe inspection) by remote-control equipment, usually a crawler inside the pipe to assess the condition of the pipe, and to check for obstructions in the line, protrusions, breaks, circumference fractures, infiltration/exfiltration, encrustation around the joints, grease, and root intrusion. In CCTV inspection, when the camera is lowered into the manhole, all other incoming pipe has to be plugged closed to restrict flow so the camera can get a clear view of the inside of the pipe. Sewer pipe can also be videotaped with minimum flow inside the pipe so long as the flow does not rise higher than the height of the camera lens, at which point the camera will become submerged under the flow.

Reasons for CCTV survey are:

Routine inspection

Troubleshooting

Compliance

Acceptance testing

CIPP (cured-in-place pipe) projects

CCTV video inspection will assess the <u>structural condition</u> of the pipe (cracks, fractures, pipe failure, collapse, deformed pipe, joint defects, lining defects, brickwork), <u>service defects</u> (deposits, roots, infiltration, obstructions), and <u>construction features</u>.

A post-repair videotape of all repair work is required to ensure that the repaired pipe has not sagged, bowed, collapsed, or become disjointed during installation and to ensure continuous flow from manhole to manhole.

Sonar CCTV- Sound Navigation and Ranging[177] is a sensing strategy which measures features of an underwater environment in which that environment transmits, reflects, and/or absorbs acoustic waves.

Sonar is normally used on large-diameter pipe and trunk sanitary sewers. The pipe cannot be plugged on either end, the flow inside the pipe is 50–100 percent, and the video camera is submerged (under water). Sonar sends out acoustic pulse from a transducer into the sewer walls to record defects in the pipe.

*Notes:* New cameras with rehabilitation technologies such as fish-eye, panoramic, zoom, sonar, and laser capabilities, and small CCTV cameras mounted on miniature tractors are replacing the much larger and bulkier cameras that have track rollers.

## 3. Repair Techniques

*External Point Repair-* This technique is used when a section of concrete pipe has collapsed, cracked, or has more than 50 percent blockage by root intrusion or debris. A point repair is used also when that section of pipe has joint displacement, when a section of pipe missing, or when the pipe is sagging from the weight of the soil above it and needs to be removed. Bypass pumping has to be set up at the upstream and downstream manholes to

---

177    http://sonar-solutions.com/html/faqs.htm

divert flows so the repair can be performed. Traffic-control measures have to be put in place to reroute vehicles from entering the construction zone. A measured section of pipe can be cut out and replaced, or the entire section can be removed. A measured section is removed by cutting it with a concrete saw. A new section of pipe will go in the place of the damaged pipe and joined with a fernco (rubber boot with metal bands) connection. Stone bedding (#57 stone) placed underneath the pipe and at the spring line supports the pipe in place from movement and settlement from compaction.

*Internal Point Repair-* This technique is used to locate and repair small to medium-size cracks in concrete pipe without having to dig down to the pipe to make the repair. A motorized camera with cable is lowered into the manhole from the CCTV truck above and pulled into position where the crack in the pipe is located. When the camera is positioned at the crack, a trolley and repair sleeve releases air and locks in place the sleeve grouting the crack. The sleeve is then pulled out after the grout has set. A post-repair videotape of the work is done to ascertain whether the repair was successful.

*CuredInPlace Pipe (CIPP)* This technique has become more popular recently within the last twenty years, but it comes with higher labor costs. It is popular because it uses trenchless technology. Trenchless means no digging or hauling of soil will occur. Bypass pumping has to be set up at the upstream and downstream manholes to divert flows so the repair can be performed. Traffic-control measures have to be put in place to reroute vehicles from entering the construction zone. The existing area remains undisturbed. The procedure uses an inverted liner bag pulled through the entry manhole down through the concrete pipe to the other end of the exit manhole. The liner bag is "cooked" at a very high temperature by one of two methods (water or air) depending on the length of the line segment. While the bag is heating up, it expands and adheres to the inside diameter surface of the concrete pipe thus creating a liner for the existing pipe, effectively lining the interior. When the cooking is completed, the liner is allowed to cool down for the proper time. The ends of the liner are cut off at each manhole with a circular hand router. If the concrete line segment has "services" (six-inch residential sewer laterals) that tie into the main line, they will be cut out with a remote crawler router with a camera and mini lights (similar to CCTV) attached and lowered into the pipe. The new liner is post-videotaped to make sure it has completely sealed to the inside pipe. The line is then tested for continuity by running water through it.

Note: If the cooking procedure is not done correctly and the bag blisters or collapses, it has failed and needs to be removed. The procedure will have to be redone correctly.

*Lateral Rehabilitation Repair-* This technique is used to remove a broken or damaged concrete connection or service lateral from the house that ties into the main line.( Note: these concrete or vitrified clay connections over time have cracked, broken, or collapsed due to settlement over the years and need to be replaced). This repair is important because it affects how sewage from residents' homes flows out to the main line. If the connection is broken at the main line, that means the sewage is not getting into the system and is spilling into the ground). The contractor will mark out the distance in the roadway on the asphalt from the upstream manhole to where the damaged connection is below ground. Bypass pumping has to be set up at the upstream and downstream manholes to divert flows so the repair can be performed. Traffic-control measures have to be put in place to reroute vehicles from entering the construction zone. Crews will saw-cut the pavement for a trench measuring eleven feet by five feet (or larger) and excavate down below the road to the depth of the damaged connection. A trench box is set in the hole over the main-line pipe and a ladder is placed inside the box for workers to enter and exit. The contractor's operator will stop digging when they feel the top of the main-line pipe and they will hand-dig around and under the main line so they can make the repair. If the contractor operator feels they are close to the pipe but cannot see it, they will use a probe rod to poke around in the soil to feel for the pipe. A bypass is set up in the main-line pipe between the two manholes where the repair will be done to stop sewage flow from entering into the bottom of the trench where they are working. The main line is saw-cut on both ends where the lateral connection ties into the main line, and the damaged connection is removed and a new section of stronger eight-inch PVC(SDR#35) pipe with a sweep wyes connection is tied in place with metal straps (bands). Before the connection is tied back in place, a push camera is sent into the service lateral that goes to the house to a distance of one foot behind the curb at resident's yard to see if the service lateral is full of roots, partially clogged with dirt or grease, or whether it has collapsed. If this condition exists, then the lateral needs to be replaced also. New six-inch PVC (SDR#35) pipe replaces the old up to a new clean-out box behind the curb in the resident's yard for the resident's convenience.

*Pipe Bursting-* This technique is used when it has been determined that a lining repair may not be used. Two pits are required (entry and exit) or upstream and downstream manholes. Traffic-control measures have to be put in place to re-route vehicles from entering the construction zone if the line segment is in the roadway. A bursting head (made of steel) is attached on the front end of a high-pressure line that bursts the existing pipe as the bursting head moves forward. The bursting head splits the old pipe and, at the same time, pulls the new pipe in place. This new pipe (HDPE) may be of equal diameter as the old pipe or larger. Pipe bursting is suitable for replacing older pipes, which are made of vitrified clay, unreinforced concrete, asbestos cement, polyvinyl chloride- PVC, and cast iron and have become brittle over time.

Advantages:

- The pipe can be replaced without disturbing the surface area.

- The work is trenchless (you don't have to dig up entire segments of the roadway).

- It reduces the amount of restoration when the work is completed.

*Notes:* Advances in static pipe-bursting systems using trenchless technology for underground utilities include water, gas, electric, petroleum, and fiber-optics piping, and they are not just limited to sewers.

*Open Cut-* This technique is used when a damaged, sagging, or collapsed pipe is below a depth of ten feet and has to be removed. Bypass pumping has to be set up at the upstream and downstream manholes to divert flows so the repair can be performed. Traffic-control measures have to be put in place to re-route vehicles from entering the construction zone. Once the exact location of the damaged pipe has been identified (usually in the roadway), a saw-cut patch is performed in the roadway longer than the length of the damaged pipe and normally five inches wide or wide enough for the size of the trench box to be lowered into the ground. Excavation then occurs down to the depth of the pipe with a backhoe or excavator. The soil material removed (which is unsuitable for filling the trench) will be loaded onto a dump truck and hauled off, and clean fill material (aggregate base) will be hauled in and stockpiled to be used later as backfill in the trench over the soon-to-be-replaced pipe. A trench box is placed over the pipe (to protect the workers) and a ladder placed into the trench to allow workers to enter and exit. The damaged pipe section (or multiple sections) is removed and replaced with a new pipe. The trench is backfilled with the stockpiled fill material.

*Jack/Bore-* This repair technique is used as a tunneling method in areas with considerable traffic and when standard open-cut excavation cannot be performed. The cutting head is a laser-guided rock-cutter-type head that will cut through tightly compacted soils and trench rock. The jack and bore method is used to bore pipe (tunnel) underneath highways, interstate roads, and roadways, and underneath and near building foundations. It is also used to bore above, beneath, or beside clusters of buried utilities, and in easements and outfalls to upsize existing sewer pipes.

Pipe sizes can vary from eight inches upward to forty-two inches or greater in diameter. Bypass pumping has to be set up at the upstream and downstream manholes to divert flows so the repair can be performed. Generally, an entry (jacking) pit and an exit (receiving) pit, typically twenty feet by forty feet, have to be excavated in front of the upstream and downstream manholes down to the depth of the existing pipe, and sections of the pipe on both ends are removed. The trench is reinforced on all four sides with metal plates thick enough to support the surrounding soils from shifting and collapsing the trench. Depending on the depth of the manhole, an engineered trench box may be required. The boring machine is lowered into the entry pit on its tracks and the boring head and tubes are set in place to start boring thru the soil. The boring head excavates the surrounding soil and the casing pipe (made from steel plates) is pulled in place (welded in sections) until it reached the exit pit on the downstream end. The sewer pipe in then inserted inside the "casing pipe" with spacers attached along the length of the pipe to hold it in place while it is being pulled through to the other end. A calculated mixture of flowable fill grout and concrete is used for filling the voids between the casing pipe and the carrier pipe as an additional stabilizer.

*Note:* During this method, the contractor is required to provide surface settlement markers, as directed by the engineer, outside the pavement areas, along the centerline of the casing, at the shoulder of the road, and at the edge of pavement (where applicable) to monitor ground and soil movement if in residential areas.

*Pipe Jacking-* This technique is considered a trenchless construction method of installing pipes or liners along a required alignment by pushing the pipes or liners into place behind the cutter head.

Advantage

- It allows for reduced surface and shallow utility disruption, flexibility in handling unexpected ground conditions, eliminating the need for two-step liners, reduced seepage into completed tunnels, remote excavation and mucking in small diameter tunnels, and increased safety as compared with traditional tunneling.

*Microtunnelling-* This repair technique is a special case of the pipe-jacking technique where the excavation process is remotely controlled and laser-guided micro tunneling (up to 1,500 feet) is generally used for constructing pipelines to a very close tolerance for line and grade. It can be used for installing pipes of any diameter up to 120 inches.

Microtunnelling can be used in a variety of ground conditions, from soft clay to rock, and up to one hundred feet below the water table without dewatering.

*Directional Drilling-* This technique is used for very tight specific trenching for shallow installation of water lines (of copper or ductile iron) from beneath the surface of the asphalt to 150 feet in length under roads, driveways, sidewalks, and sensitive landscaping areas where traditional digging would be cumbersome and labor-intensive.

Directional (horizontal) drilling equipment can be small in size or large in size depending on the scope of the job.

*Sewer Pipe Installation-* Sewer pipe can be replaced in sections (called a point repair), or the pipe can be replaced in entire lengths. In a point repair, it has been determined that the entire point does not need to be replaced but only a section of the pipe that has been cracked, broken, or disjointed from a connecting pipe. If the entire section of pipe (normally twenty-foot lengths) has to be replaced, it will be removed from the connecting pipe on both ends. The old pipe will be dug up, rock bedding with #57 stone will support the new pipe, and the pipe is set in place using a laser level to determine pitch (fall) with the existing pipe. Stone bedding (#57) of one to two feet will be placed over the new pipe as protection. Existing soil that has been removed from the trench can be used as backfill and compacted if it has been determined that the soil is suitable for compaction. If the soil is not deemed suitable for compaction (by testing), then it will be hauled off and new clean fill material will be used in its place.

Notes: Individual pipes are joined together at the bell ends. Pipe lubricant is applied to act as a seal around the entire pipe at both ends as they are being connected. Every joint of pipe in place is tested from manhole to manhole to verify there are no leaks at the joints. If the joint test fails a particular section of pipe, the cause of the leak has to be found and corrected before the test will pass. All pipe-joint sections have to pass.

*Manhole Installation-* A new precast manhole will replace an older brick structure when it has been determined through inspection that: a) the walls inside the manhole are leaking (grout is missing between the bricks); b) the invert (bottom of the manhole) has washed out and water is coming up from the bottom into the structure; c) and the manhole is collapsing. The incoming and outgoing pipes have to be bypassed so the flow does not go into the manhole while it is being replaced. The old structure will be demolished and removed from the site. The new manhole location will be marked out on the ground with spray paint. Stone will be placed on top of the ground and compacted to act as a base for the structure. The base structure will be set in place and riser sections (two to four feet) added until the desired height to grade is achieved. The outside and inside joints have to be sealed to prevent leakage. A new invert and bench have to be installed on the bottom of the manhole so the flow can pass through and not build up inside the manhole.

*Manhole Core-* This repair technique is performed using a coring machine with a circular drill attachment to core into a brick manhole or precast manhole without compromising its structural integrity to install a pipe. The pipe going into the manhole through the core must have a rubber boot with tightened bands and be concrete-sealed to prevent leakage.

*Manhole Raise-* This repair technique is performed when it has been determined that the manhole is buried below the finish grade and it cannot be accessed. The manhole structure has to be raised to grade level by adding several new courses of brick, and a new ring and cover must be installed to replace the old ones. Manholes that are buried can be determined several ways by using a metal detector above ground in the roadway or with a CCTV camera that has located the top inside the manhole while doing routine inspection. Buried manholes can be in homeowners front yards or in the sides and backyards of their properties, covered over with sod or dirt. They can be buried in easement (off the right-of-way) areas, near creeks, and by railroad tracks, or paved over by asphalt crews.

*Manhole Rehabilitation-* Due to old age or poor construction, manholes will need repair. This repair technique is used when it has been determined that the existing manhole can be kept but will require rehabilitation to put it back in good serviceable condition. The inside brick walls (if the manhole is of brick structure) will need to be sealed in between the joints to stop water penetration. The invert bottom will be sealed along with the bench (manhole bottom) to stop seal ground water penetration from underneath. The inside walls (if the manhole is a precast structure) will need to be sealed between the joints or cracks in the wall with leak stop. The walls can then be lined with a spray-on coating up to a three-inch thickness (fiber-reinforced cementitious liner) from the invert up to the ring or with a 100 percent solid, corrosion-resistant epoxy coating.

*Stream Crossings-* This is one of two techniques used to replace and install new sanitary sewer piping crossing perpendicular to a stream. Bypass pumping has to be put in place to keep the flow going between manholes. The existing sanitary sewer piping has to remain in place and active until the new sewer pipe is installed and tested. To prevent erosion at the stream bank, the soil will be stabilized by covering it with filter fabric and type I (large) or type III (medium) rip rap.[178] To cross the creek, railroad ties are placed across the creek using machinery; large corrugated pipes (normally sixty-inch corrugated metal pipe) are placed in the stream where the natural flow of the creek passes through these pipes. Medium riprap is placed over the top of this pipe to walk on. Once the permanent sewer pipe has been installed across the creek, all this temporary work is removed.

*Aerial Stream Crossings-* This is the second technique used to replace and install new sanitary sewer piping crossing perpendicular to a stream. The sanitary sewer piping will be installed above the stream on concrete piers. Oftentimes, as you track down manholes in outfall areas, the sewer piping will go across the stream to another manhole. At the banks of the creek or stream, the concrete piers and brackets supporting the sewer pipe have severely eroded away over time and may need retrofitting, or they have completely washed out, collapsing the sewer line in severe cases from bank erosion. The manhole and the sewer piping have collapsed into the creek and will need to be completely replaced. Bypass pumping has to be put in place to keep the flow going. The old sewer pipe has to be removed and new sewer piping installed (inside the steel casing across the creek). Depending on the width of the stream, concrete abutments may be installed on either

---

178    Stones or chunks of concrete thrown together without order along a stream bank to control erosion

side of the stream to receive the casing pipe before going into the manhole. To prevent erosion at the stream bank, the soil will be stabilized by covering it with filter fabric and type I (large) or type III (medium) riprap.

*Water Line Installation-* Water lines that are old, aging, corroded, leaking, or too small and restricting flow need to be replaced. Water lines go from hydrant to hydrant usually in the street or under the sidewalk, depending on the code ordinance, and on the opposite side from the gas line. Water lines with low flow are a life-safety issue and need to be addressed. Upsizing water lines to eight inches are ideal in residential areas with a working pressure of 250 psi. The pipe reduces from the main line with a "tee" and gate valve (control) to connect to the fire hydrant. A "corporation stop" ties from the main line to each residential homeowner's meter box behind the curb and controls the velocity of the water from the main line to the meter.

*Fire Hydrant Installation-* Fire hydrants need to be maintained on a routine basis. Fire departments, as part of their duties, will check fire hydrants for flow and pressure to make sure they are in good working order. The fire hydrant may have parts internally that have been damaged. To replace the hydrant, the flange bolts have to be removed to get the hydrant off from the stem.

*Notes:*

Fire hydrants that are leaking, that have low flow, or that have been hit and damaged need replacing. The stem connects the hydrant to a ninety-degree bend (below grade), and then a straight section of pipe ties that to the main line. A thrust block (of concrete) with rods holds the pipe in place at the bend for thrust.

Fire hydrants on the grounds of apartment complexes and on private property are the responsibility of the owner to maintain. Check with your complex to see if hydrants are routinely maintained and are operating with full capacity. It would be a shame if the fire department connected to the hydrant and there was no water.

Many cities are replacing fire hydrants and upsizing water line piping because of low pressure.

## Contractor General Maintenance

City of Atlanta specifications state that the contractor shall furnish and install materials, equipment, and labor that is reasonable and properly inferable and necessary for the proper completion of the work, whether specifically indicated in the contract documents or not.

The contractor shall perform the work complete, in place, and ready for continuous service.

When contract crews are working on SSES projects, they will turn in a daily construction schedule the morning of that day of work to the project manager for the city. Inspectors representing the city will have a critical inspection check sheet with them upon arriving at a work site to record and verify that the contractor has certain notifications, signs, and permits on hand. Although inspectors do not control the contactor's means and methods, they are in place as the "eyes and ears," of the city, or they are the first line of defense for problems in the field. Inspectors report to the resident engineer or project manager when the need arises for an engineering decision. Inspectors will record the day's activities on their daily reports. They can possess various certifications needed in the field, including CCTV, Occupational Safety and Health Administration (OSHA) Competent Person, trenching and excavation, and erosion control.

Inspectors confirm that the contractors and their crews are performing work as prescribed in the city's specifications.

Contractors working under the watershed-management program are required by their contracts as part of their scope of work to perform maintenance, such as flushing (jet cleaning), clearing pipes of debris, and vacuuming out (by suction) debris in lines so that CCTV video recordings of line segments can be performed to assess the condition of the pipe and manholes.

Also, the bases of sewer pipes that cross creeks or run parallel to creeks are vulnerable to being washed away, causing the pipes to become disjointed and sag. Often these lines are at the lowest point and are susceptible to being clogged with all types of debris, silt, and roots. Manholes and pipelines in the outfall areas are just as vulnerable because they are out of site but need to be maintained. Proper maintenance would be to keep the structures vacuumed out regularly so they do not become permanently clogged.

Maintenance directors in the public works department and/or in the watershed-management department should have set criteria for allocating funding for repairs from reports they receive from their area supervisors (administrators) in the field. This allows the directors to get a more accurate assessment and give this information to the deputy commissioner over the production of wastewater for review.

A good maintenance program will assess a cleaning of the sewer system every five years with "hot spots" more frequently. Mitch Flanders was a City of Atlanta project manager for Contract D in the west area SSES contract. He stated that a good maintenance program includes periodic assessment of pipe and manhole conditions, scheduling maintenance and repairs logistically to account for traffic or environmental conditions, ongoing training for personnel, and regular preventive maintenance on equipment (both survey and repair equipment should include calibration and measuring). Marketing and research should be done to assess new and better methods along with equipment.

In an interview with a project manager who managed the contractor crews for SSES Contract D, he stated that a good maintenance program includes early detection of root intrusion and early root removal in the main line in areas such as near creeks. Also included are outfall areas, easement (off the road) areas high-forested areas, and areas where vibration from railroad tracks reaches the underground pipe. In these areas, the pipes can become dislodged or they can crack, and crews must check the areas where the pipe has to go around the creek in lieu of going underneath it. Another part of maintenance is replacing missing rings and covers. Often when crews are doing recon or CCTV observation in easement areas or in outfall areas, they will come across open manholes with missing rings and covers. This is a safety issue, with the risk that people, animals, and objects can fall into the manholes.

*Notes:*

The contractor is responsible for contacting all utility companies that have utilities within or adjacent to all work areas within the right-of-way of construction activities and to coordinate with them to have these utilities relocated if they are identified as being in the way.

The contractor has to mark and maintain monuments and benchmarks located within the areas of work.

## Buffer Zones

In steams or creeks, buffer zones are measured horizontally from the point where vegetation has rested by normal stream flow or wave action—twenty-five feet for warm-water streams and fifty feet for trout (cold) streams. These distances serve as "safe zones" for the protection of the stream and the ecosystem in the stream. The function of buffers is to reduce runoff velocities, to act as a screen for visual pollution, to reduce construction noise, to improve aesthetics on the disturbed land, to filter and increase infiltration of runoff, and to cool the rivers and streams by providing shade.

Buffer zones are protected by installing silt fencing along the disturbed construction area(s) near the creek.

## Bypass Pumping

Bypass pumping is used when manholes or sewer pipes need to be repaired or replaced. The procedure is to put a blow-up plug in the outgoing pipe, a suction hose (size varies) that drops inside the manhole to the incoming invert[179] and sucks the flow from the bottom of the manhole up to and out the pump through the discharge hose and then down to the next open manhole where the flow is released (thus the name bypass). The aim is to stop any flow coming into the area that needs to be repaired or replaced.

*Note:* As a precaution, a second, backup bypass pump should be primed and ready in case the primary pump goes out, which does happen from time to time.

## Cleaning

Contractors are required, in inspecting the conditions of manholes and pipes prior to any repairs, to assess if any obstructions are blocking or re-stricting the flow in pipes getting from manhole to manhole. One piece of equipment contractors have in their arsenal is a vacuum truck.

This truck has very powerful suction and water-jet spraying abilities. It has a hydraulically operated, front-mounted hose reel (one thousand feet long) with a high-capacity water pump to jet the sewer line clean, a reservoir tank that hold a minimum one thousand to 1,200 gallons of non-potable[180] water, a suction nozzle (hydraulic boom) on the top of the truck that can rotate up to 270 degrees with suction tubes normally six inches diameter in

---

179    Inside bottom of the pipe

180    Water that has not been examined, properly treated, and not approved by appropriate authorities as being safe for consumption.

ten-foot sections stored on the truck, and a separator tank that contains the debris sucked out from the pipes, manholes, storm drains, or catch basins.

The purpose of the vacuum truck (or vac truck) is to suck out any objects such as trash, mud, grease, bricks, concrete, rocks, asphalt, silt, sewage, water, sanitary products, plastics, paper, rags, dead rodents, metal, or any debris stuck inside the pipe or manhole that are lodged or impeding the flow inside the pipes or the manhole.

Once the pipe has been repaired or replaced, it will get a final cleaning with the jet hose, and CCTV inspection will complete the repair.

*Notes:* A smaller piece of equipment called a "prowler easement cleaner" is used when the larger truck cannot be driven to a manhole(s) in areas such as a near a creek or a stream, off the right of way, on residential or commercial property, and in easement areas with low to medium brush that can be cleared to access the manholes.

## Clearing

Oftentimes sewer piping has to be repaired or replaced in outfall areas for upsizing (as in the case with sewer relief) where there are trees, kudzu, vines, and overgrown vegetation. These areas will need clearing to install the new sewer piping according to the design drawings and specifications.

Contractors hire companies that specialize in cutting down trees, clearing brush, and chipping the brush into mulch. The company has a foreman (usually a certified arborist who can determine the species and health of the tree(s) and determine whether they should be removed or can be removed without destroying surrounding trees. Normally, the city arborist has already gone to the site and determined by marking the trees that are in the right of way of the sewer line and therefore need to be removed.

The contractor has to strip all vegetation from the surface that are within the right of way of construction activities to including disposing of it and, where necessary, replacing/restoring areas with seed and straw or the same plantings in residential areas. Some plans call for planting new trees.

All materials deemed unsuitable(contaminated roots or topsoil) and not reusable have to be hauled off and disposed of properly and legally.

There are several ways contractors remove trees from site.

One way is to climb the tree (with spiked shoes, and rope harness). The climber will then have a hand chain saw pulled up to him for and cutting the limbs and branches away from the body starting from the ground until they reach a point just before the top of the tree. Then the contractor cuts the main portion of the tree in sections (cutting his or her way down, in reverse) all the way back to the stump. The sections of cut-up tree are loaded onto a log truck and hauled off, while the branches are shredded and used as mulch.

The other way is using a Shinn cutter machine (a modified Komatsu PC228 LC zero turn) track excavator that has a cage built around the windows and door and on top of the cabin where the operator sits. The operator uses the cutting attachment at the end of the machine's front arm to completely grind off the tree limbs, the branches, and the main body of the tree down to the stump. The Shinn cutter can also cut down overgrown vegetation, much like a lawn mower.

## Testing

As part of the Sanitary Sewer Evaluation Survey testing of various types by qualified contractors, testing is performed to inspect the structural integrity of the sanitary sewer system and address capacity and maintenance issues that undermine the performance of the system.

The most commonly and frequently used tests in the field for assessing and determining problems to the sanitary sewer system are the following:

Smoke Test-

Smoke tests are conducted as part of the evaluations in areas that are suspected to have inflow problems to find leaks and defects. Smoke testing is an economical and relatively fast method for identifying the location(s) of inflow sources, such as structural foundation drains, yard drains, storm sewers, and illegal connections. For this test, residents are asked not to be home during this process in the event smoke backs up into the resident's home. Windows should be partially open also. The area of smoke testing should be blocked off during testing. Each test will consist of two sections of pipeline, generally 600–800 linear feet of eight-inch to twelve-inch pipe. A portable air blower is set over the manhole in between the two test lines in either direction, with sandbags on the bottom of one manhole and a stopper (line plug) on the bottom of the other to help confine the smoke in specific sections of the line. Smoke is forced into a gravity sewer line using the blower at high pressure. If there are any holes, cracks, or other defects in the sewer line, smoke will seep through and visually identify the problem areas. The smoke is nontoxic, has a slight odor, and is white to gray in color.

*Note:*

1) It is best to perform this procedure on a dry day when water is not leaking into the line.

2) It is not uncommon to see smoke coming from cracks in the asphalt paving, showing points of surface water entry.

Dye Test-

Dye tests are critical to identifing rain or ground water entry points into the sanitary sewer system, using colored dye poured into residents' toilets that flush out to the sewer laterals (from homes). Roof drain leaders, driveway drains, and area drains are key to locating where a pipe drains to or if sanitary sewer and storm-drain catch basins are connected to the sanitary sewer system. The dye is nontoxic and can come in a variety of colors, with the most commonly used being fluorescent green and yellow. In downspouts, a powdered dye is inserted and this causes the water in that pipe to become fluorescent green and identify where the pipe is emitting its water in downstream manholes.

Note: Residents should be given a 1-week notice by letter prior to any testing.

Dye-water flooding is performed with CCTV to confirm potential inflow and infiltration (I/I) in a storm-drain system. Plugs are inserted in the invert of manholes (to isolate line segments), the line is flooded (surcharged) to street level to simulate a storm flood to see where there are defects or blockages in the system. The rate of infiltration can also be noted. Knowing where the leaks are located in a system will assist in determining the most cost-effective means to repairs.

## Other Testing

From time to time during the process of the work, the engineer may require that testing be performed to determine that materials used in construction meet the specification requirements. Nothing that is put in the ground for infrastructure work is left without doing a test or without the structure being tested. Tests determine whether the material(s) installed can withstand the pressure that will be put on it. Tests show that the backfill material or existing soil material will hold up under the pressure that will be put on it. Tests show that the water lines and water meters will hold up under the pressure put on them. Air and vacuum tests show that manholes will withstand the pressure put on them without leaks.

Testing will include, but is not limited to:

Low Pressure Air Test- performed on newly installed sewer pipes from the manhole-to-manhole segment, four pounds per square inch (psi) for four seconds (depending on the diameter size of the pipe; refer to specification chart).

Pressure Test- performed on manholes with a vacuum test.

Pnuematic Test- for testing five pounds per square inch (psi) of air pressure for five minutes.

Hydrostatic Test- for testing water-line pressure.

Compaction Test- for testing soils after backfilling trench.

Concrete Test- to determine proper temperature, slump, and proper strenght (psi) after it has hardened.

Aggregate Test- to determine proper mix and size of stone.

Steel and Welding Tests- to determine the strenght of the beam and proper amount of tap weld.

Joint Testing- to determine if the connected pipe will hold up to static pressure.

Bituminous Asphalt- check for the proper temperature during spreading, proper thickness applied, and strenght after cooling.

# Part 6 - Permits and Permitting Work

## General Permits

Under the general conditions for permits in Georgia, the contractor shall comply with all applicable state laws, municipal ordinances, and the rules and regulations of all authorities with jurisdiction over construction of the project and all work performed within the right of way[181] of GDOT, and shall be in accordance with GDOT regulations.

The contractor has to apply and secure all other required permits and inspections outside what the City has secured and provides as the building permit for the project to include:

Traffic control, utility encroachment, haul routes, street closures, noise, blasting, NPDES storm-water discharges, notice of intent (NOI) permit.

Permits are required for the execution of all city work. Permits allow for tracking construction activities. Contractors need construction permits (utlity) to perform contract work in Atlanta's right-of-way and easement areas. Permits give the contractor, neighbors, and the city (the issuing authority) the assurance that the minimum standards are met in constructing and repairing construction work. Permits allow contractors authorization to enter right-of-way areas, personal property, streams, creeks, railroad crossings, and state roads, and they identify who is responsible in the event

---

181     In the road

of prosecution. The standards set forth on the permit protect the integrity of the work, the welfare of the public, and the safety of the contractor.

## Other types of permits contractors would have to apply for with the City of Atlanta include:

*Blanket Traffic Permit*- allows contractors to perform construction operations on City of Atlanta streets.

*Noise Ordinance Permit*- allows contractors to perform construction operations after normal working hours.

*Parking Meter Permit*- allows contractors to perform construction operations in front of City of Atlanta parking meters.

*Georgia DOT Permit*- allows contractors to perform construction operations on the state's right-of-way areas.

*Road Closure Permit*- allows contractors to close off streets/roads to perform construction operations on City of Atlanta streets.

*CSX Permit*- allows contractors to perform construction operations near CSX railway.

*Utilities Permit*- acts as a blanket traffic permit with rehab construction conracts.

*MARTA Permit*- allows contractors to encroach on MARTA property in order to perform construction operations.

*County Permit*- allows contractors to enter and work on county-owned property.

*Tree Ordinances*- allows clearing of trees located inside/outside of permanent easement for the purpose of building an access road.

*Army Corp of Engineers (Nationwide Permit 3)*- allows contractors encroachment for maintenance of existing utilities in US waters.

*Erosion Control (State): Land Disturbance-* needed if there is more than one acre of disturbance or if construction is within 200 feet of a creek or an intermittent stream, such as a ditch.

# Part 7 - How Everything Works

## Combined Sewer Overflow (CSO)

Combined sewer overflows are designed to collect rainwater runoff, domestic sewage, and industrial waste all in the same pipe. Most of the time, combined sewer systems transport all of their wastewater to a sewage-treatment plant, where it is treated and then discharged to a water body (such as the Chattahoochee River in Atlanta). However, during periods of heavy rainfall or snowmelt, the wastewater volume in a CSO can exceed the capacity of the sewer system or treatment plant.[182]

These overflows, called combined sewer overflows (CSOs), contain not only storm water but also untreated human and industrial waste, toxic materials, and debris. They are a major water-pollution concern for the approximately 772 cities in the United States that have combined sewer systems. CSOs are discharges from a municipality's wastewater conveyance infrastructure that are caused by precipitation events such as rainfall or heavy snowmelt.

Under the Clean Water Atlanta program, the federal EPA and state EPD authorized the city to implement its plan to eliminate water-quality violations from combined sewer overflows (CSOs).[183] The city's plan involved a combination of tunnels and separation of selected sewer areas. The city submitted a refined plan to the EPA and the EPD that would increase

---

182    http://cfpub.epa.gov/npdes/home.cfm?program_id=5

183    http://cleanwateratlanta.org/ConsentDecree/Elements/CSO.htm

the water-quality benefits of proposed portions of the plan and reduce the lengths of the proposed CSO tunnels.

## Sanitary Sewer Overflow (SSO)

The EPA states that unintentional discharges of raw sewage from municipal sanitary sewers occur in almost every system. These types of discharges are called Sanitary Sewer Overflows (SSOs). The untreated sewage from these overflows can contaminate our waters, causing serious water-quality problems.[184]

Under the Clean Water Atlanta program, beyond the nineteen-square-mile combined sewer area, Atlanta's sewer is separated (85 percent of the total sewer area).[185] Sanitary wastewater flows in its own pipe to the treatment facility, and storm water flows in a separate pipe to the receiving stream. During SSOs, a mixture of untreated sewage, groundwater, and storm water overflows from the pipes or from manholes connected to the pipes. Many sewer lines run alongside creeks and streams, many of which are adjacent to private property.

In order to address improvements in sanitary sewer systems, the city, the EPA, and the EPD negotiated a second comprehensive settlement, titled the First Amended Consent Decree (FACD).

The FACD prescribes evaluation and improvement measures to eliminate SSOs and to protect the city's investment in upgraded water reclamation center (WRCs). The SSO program included the development and implementation of Maintenance, operations, and management (MOM) programs, completion of the city's capital improvement program (CIP) for the sewer system, an aggressive grease-management program, and the evaluation and rehabilitation of existing sewers.

---

184    www.epa.gov

185    http://cleanwateratlanta.org/consentdecree/Elements/SSO.htm

The EPA states problems that can cause chronic SSOs include:

- *Infiltration and Inflow (I&I)*: too much rainfall or snowmelt infiltrating through the ground into leaky sanitary sewers not designed to hold rainfall or to drain property and excess water inflowing through roof drains connected to sewer-service lines.

- *Undersized System:* sewers and pumps are too small to carry sewage from newly developed subdivisions or commercial areas.

- *Pipe Failures*: blocked, broken, or cracked pipes; tree roots grow into the sewer; sections of pipe settle or shift so that pipe joints no longer match; and sediment and other materials build up, causing pipes to break or collapse.

- *Equipment Failures:* pump failures; power failures.

- *Sewer Service Connections:* discharges occur at sewer-service connections to houses and other buildings; some cities estimate that as much as 60 percent of overflows come from service lines.

- *Deteriorating Sewer System:* improper installation; improper maintenance; widespread problems can be expensive to fix and develop over time; some municipalities have found severe problems, necessitating billion-dollar correction programs; often communities have to curtail new development until problems are corrected or system capacity is increased.

## Water Reclamation Centers:

The functions of water-reclamation centers (WRCs) are for treating wastewater within the city. There are four water-reclamation centers in the City of Atlanta. All have been upgraded or completely refurbished.

Note: Information provided by the cleanwateratlanta.org

1) R.M. Clayton- located in northwest Atlanta, expansion and improvements were completed to eliminate the bypasses, spills, and other problems that had affected the plant's operations over the years. New sand filters were installed to keep phosphorus levels within or below state standards. State-of-the-art odor-control technology was installed to reduce odors disseminating throughout surrounding areas. A chlorination/dechlorination disinfection system was installed to kill microscopic bacteria. In addition, treatment capacity was expanded from one hundred million gallons per day (mgd) to 122 mgd. Additionally, in accordance with the new clean- air regulations, improvements were made to the plant's incinerator.

2) Utoy Creek- located in southwest Atlanta, expansion and improvements included new sand filters to lower the level of phosphorus in treated wastewater. This element of phosphorus control enables the facility to meet state permit requirements. A chlorination/dechlorination disinfection system and odor-control systems were installed to meet current and future demands. The treatment capacity was expanded from 36 mgd to 44 mgd. Sludge, the solid by-product of the sewage treatment process, is incinerated and, less frequently, landfilled. Future plans include installing a new heat-dryer technology to treat digested primary and waste-activated sludge to produce a marketable soil conditioning fertilizer.

The treatment process removes biological and other impurities from the wastewater, making it suitable for discharge to the Chattahoochee River in three phases: primary treatment, secondary treatment, and tertiary treatment.[186]

*The primary treatment phase* begins with the removal of floatable and settleable solids. Removed solids are hauled to a landfill for disposal.

*The secondary (biological) treatment phase* reduces the concentration of organic and chemical pollutants.

*The tertiary treatment phase* removes the remaining suspended solids, disinfects the treated water using ultraviolet radiation, and increases the dissolved oxygen concentration before discharging to the Chattahoochee River.

3) South River- located in southeast Atlanta, expansion and improvements included new sand filters that were installed to keep phosphorus levels within or below state standards. State-of-the-art odor-control technology was installed to reduce odors disseminating throughout surrounding areas. A chlorination/dechlorination disinfection system was installed to kill microscopic bacteria.

4) Intrenchment Creek- located in southeast Atlanta/Dekalb County, it has a smaller capacity compared with the other three facilities. Expansion and improvements included upgrading of the effluent pump station to increase its capacity to transfer treated wastewater to the South River WRC for future processing during wet-weather conditions. Standby diesel generators were installed for use during emergencies. High- and medium-voltage electrical and instrumentation systems, including the plant's control computers, were modernized to improve the

---

186    http://www.atlantawatershed.org/facilities/wrc/utoy.htm

operational reliability and efficiency of the WRC. A reuse-water system was installed to enable treated effluent to be utilized for wash-down and other nonpotable uses. The WRC's influent-screening equipment and digester gas-handling systems were rehabilitated.

## Pumping Stations:

The function of pumping stations is to collect and pump wastewater from the sanitary and combined sewer collection network through large-diameter pipes, pumps, valves, grit chambers, and gas monitors to the City's four water-reclamation centers. New technology has been added to the system, such as new telemetry panels[187] that transmit data to a monitoring center, thereby eliminating the need for manual facility inspection two or three times daily and eliminating fast response by a trained crew when needed. This improved system will decrease the incidence of sewage overflow and protect groundwater and area streams.

There are ten pumping stations throughout the city:

Hartsfield-Jackson Repump Station

Northside Tank and Repump Station

Adamsville Tank and Repump Station

Long Island Creek Booster Station

Chattahoochee River Intake

Mt. Paran Booster Station

Welcome All Booster Station

West Wieuca Booster Station

---

187    http:www.cleanwateratlanta.org/consentdecree/Elements/PumpStations.htm

Peachtree-Dunwoody Booster Station

South River Pumping Station

## Tunnel Pump Stations:

They pump the excess flow from the tunnels to existing water-reclamation facilities for treatment during low-flow periods.

## Water Treatment Plant:

It is a facility that collects, processes, and distributes potable (drinking) water through a distribution system (pipes, valves, pumps, tanks, filters, screens, cleaning, and chlorination processing plants) for public use.

There are three water treatment plants that service the City of Atlanta and process 100 percent of Atlanta's drinking water from the Chattahoochee River. They serve as Atlanta's primary water source, supplying water to retail, residential, commercial, and industrial customers within the city and to portions of Fulton County south of the Chattahoochee River and in South Fulton County. The Chattahoochee Hemphill Water Treatment Plants address a variety of activities at the two complexes, including:

- Cleaning, inspecting, and renovating filters
- Valve and actuator replacements and modifications
- Piping and pump replacements
- Weir replacements at sedimentation basin header flume
- Filter building, operator lab station replacement
- Chemical and filter-building improvements and roofing replacement
- Demolition and replacement of dehumidification equipment

- Electrical equipment replacement
- Safety and security improvements

The Chattahoochee Water Treatment Plant- processes 65.9 mgd (million gallons per day).

The Hemphill Water Treatment Plant- processes 136.5 mgd (million gallons per day). It is a manually operated water-treatment plant that is staffed twenty-four hours a day, seven days per week.

North Area Water Treatment Plant (Atlanta-Fulton County). It is jointly owned by the City of Atlanta and Fulton County and is operated by the Atlanta-Fulton County Water Resources Commission.

## Wastewater Treatment Plant:

It is a facility that collects and processes wastewater from residential, industrial, and commercial customers through a distribution system (pipes, valves, pumps, tunnels) into interceptor pipes, which then channel the wastewater into the sewer main trunk that leads to the treatment plant.

Once the flow (millions of gallons per day) reaches the plant, it goes through large bar screens that remove the trash and large items that you do not want going through the plant. Next, the flow goes through the effluent pump house, the grit removal system, the primary basins, onto the scum hopper, the solids sludge hopper, the aeration basin tank, and into the clarifier (sweep) that brings the sludge up and into the digester that receives the solids, breaks the solids down by high heat, and pumps them into the centrifuge. The solids are pulled out and hauled off to the landfill. Lastly, the flow is sent into a UV (ultraviolet) chamber to disinfect the wastewater from the clarifier, and the cleaned water is released back into the river.

*Note:* Testing of the released water is performed daily to determine water quality and whether the water meets EPA standards.

## Reservoirs:

A reservoir is any natural or artificial holding area used to store, regulate, or control water prior to distribution into homes and businesses.

There are two reservoirs located at the Hemphill WTP that supply water to the plant.

The reservoirs hold rainwater during dry days and the water is pumped out of the reservoir from a pump station through underground, large-diameter piping (twenty inches and greater). The pumping process uses a series of strategically placed blow-off valves and booster pump valves along the line to get the raw water to the water-treatment plant for processing and disinfection to turn it into drinking water.

Hemphill Raw Water Reservoir #1- has a storage capacity of 180 million gallons.

Hemphill Raw Water Reservoir #2- has a storage capacity of 345 million gallons and is the sole source of raw water for the Hemphill Water Treatment Plant.

## Chattahoochee River Intake Facility:

Water from the Chattahoochee River Intake enters and flows by gravity to the Chattahoochee Raw Water Pumping Station. This pump station was constructed to replace the old stream station. Water is screened and chemicals are applied to it before it leaves the intake. The water is sent from the intake to the Chattahoochee Water Treatment Plant (WTP) by the Raw-water pumping facility. The pump station is totally electric and is manually operated. It is staffed twenty-four hours a day, seven days a week. It consists of five service pumps that send water to the Hemphill reservoirs and four low-service pumps that send water to the Chattahoochee WTP on demand.

## Trunk Relief Sewers:

During wet weather (rains and flooding), flow in the pipes can exceed capacity, causing the manholes to fill up quickly and surge in combined sewers. Trunk-relief mains carry large volumes of sewage over long distances.

The City of Atlanta plans to upsize much of its medium-diameter pipe to larger-size pipe in the South River and Clear Creek easement and outfall areas. Pipes that are twenty-four" and thirty inches will be upsized to thirty-six and forty-two inches. This is called sewer relief. These upsized pipes are installed to carry flow in excess of the capacity of an existing sewer from a district in which the existing sewers are of insufficient capacity. The sewer piping is installed to supplement and/or alleviate the system of surging manholes.

## Recycling and Materials Management:

Over the years, Atlanta has made great strides in encouraging more recycling among its half million inhabitants.[188] Initiated by the Georgia Comprehensive Solid Waste Management Act of 1990, Atlanta developed its own Solid Waste Management Plan in 1995 to address the city's waste stream, from waste reduction and collection to disposal. Sustainable Atlanta established the Recycling and Materials Management (RMM) Task Force, with active members from various recycling businesses, organizations, and institutions within the city.

## Transmission Mains:

Transmission mains satisfy water demand in carrying water throughout the city in large-diameter piping to the water-treatment facilities. There are five transmission mains throughout the city.

---

188    www.sustainableatlanta.org

They are Hemphill, Southwest, Fairburn Road, Koweta Road, and Adamsville.

## Storage and Treatment Facilities:

Custer CSO Storage and Treatment Facility[189] is a new ten-million gallon, linear storage facility, in conjunction with the existing thirty-four-million-gallon Intrenchment Creek CSO Storage Tunnel, that was constructed to increase the total storage capacity to forty-four million gallons. A thirty-foot-diameter vertical construction access shaft, approximately 120 feet deep, was excavated into rock and its walls reinforced with concrete where required for stability. This shaft will provide access to construct the below-ground storage facility that was excavated into solid rock. In addition, the existing East Area CSO treatment facility was upgraded to treat the CSOs to a higher level. Before renovation, the current flows pumped into the tunnel were treated using bar screens (to trap small, medium, and large debris), grit removal, settling basins, and chlorine disinfection tanks/equipment. The upgrade to the facility will include the addition of fine screening, filters, and enhanced disinfection systems to control harmful bacteria and other residuals. This storage and treatment facility will store and transfer combined sewage to the South River treatment plant. Storage is needed because it has been determined that after the CSO problems are corrected, the peak wet-weather flows would exceed the flow that the plants can treat without violating the discharge permit. Storage would provide equalization of the peak wet-weather flows.

## Tunnels:

Tunnels carry sewage and storm runoff in large-diameter piping from one point to another ending at the wastewater-treatment plant. The design of

---

189    http://wwww.cleanwateratlanta.org/CSOTunnels/custerave/default.htm

relief tunnels was intended to decrease sewage overflows and storm runoff through the utilization of new and larger lines during dry periods when the treatment plant can properly digest all the flow coming into the plant.

Tunnels are needed because after heavy rainfall, the combined sewer system, which is located in the core of the city, exceeds flows causing both storm water and wastewater to flow from a manhole onto the surface or into streams.[190]

The primary piece of equipment used in tunneling large diameter is the Tunnel Boring Machine (TBM) used to bore through solid soils and hard rock.

*Note: Atlanta was already served by two wastewater tunnels—Three Rivers and Intrenchment Creek.*

The New West Area Combined Sewer Overflow Tunnel[191]- was constructed below the ground in bedrock. The West Area CSO Storage Tunnel will capture, store, and convey CSO from the Clear Creek, Tanyard, and North Avenue CSO basins. The storage tunnel will be approximately 8.5 miles long with a twenty-four-foot finished diameter and will store up to 177 million gallons of overflow from a rainstorm. When the rainstorm is over, the CSO will be conveyed to a dedicated CSO treatment plant for pollutant removal and ultraviolet (UV) disinfection before being discharged to receiving waters. A new submersible pump station with a capacity of eighty-five million gallons per day (mgd) was constructed at the end of the tunnel (at the R.M. Clayton Center) to lift the stored flow from the tunnel treatment at the new dedicated CSO treatment plant. The pump station was sized to allow a full tunnel to be emptied within a two-day period.

---

190    www.kiewit.com/about-us/˜/media/assets/kieways/2011/2011_06701k.ashx

191    http://www.cleanwateratlanta.org/CSOTunnels/west/default.htm

Orne Street sewer relief tunnel, Phase III- (under the capital improvements project)

Nancy Creek Tunnel- reduced SSOs in the North Atlanta/Dunwoody area by 70 percent.

West Area CSO Tunnel- the 8.5-mile, twenty-four-foot-diameter tunnel can store up to 177 million gallons of combined sewage for transfer to a dedicated treatment plant. The city replaced about one hundred miles of aging water mains that were leaking.

East Area CSO Tunnel- an extension of an existing tunnel, it will convey flows from the east area of Atlanta to the Intrenchment Creek CSO facility for treatment before discharge into Intrenchment Creek in DeKalb County.

South River (two-mile tunnel) will include two construction shafts. The shafts will be excavated into the rock and its walls will be reinforced with concrete and rebar for stability. These shafts will provide access to construct the tunnel, two intake structures, and drop shafts. Drop shafts and chambers will transfer flows from the existing tunnel and 1,200 feet of connecting tunnels. The tunnel spans nine thousand feet with a fourteen-foot finished diameter from the South River WRC to Macon Drive at the South River Bridge.

## Load Control Center:

The city's Load Control Center uses state-of-the-art computer systems to monitor and control the systems at the water-treatment plants and river intake facilities at Chattahoochee and Hemphill. There are two Settled Solid facilities to service the City of Atlanta at the Chattahoochee and Hemphill water-treatment plants. They process all the solid waste from the treatment plants.

## Lift Station:

It is the primary collection facility for wastewater within the gravity-flow portion of a wastewater collection system. Mechanical pumps feed the wastewater through force mains (pipes that transport wastewater under pressure from a lift station) to the wastewater-treatment facility for processing.

## Utoy Creek Administration Building/Laboratory:

This building and analytical laboratory at the Utoy Creek WRC facility is thirty thousand to forty-two thousand square feet and also serves as a visitor center. The building has a cafeteria, a seventy-five-seat auditorium, meeting rooms, and a teaching laboratory—all available for use by the surrounding community and by tour groups wanting to learn about water-treatment facilities (by scheduling).

*Note:*

Access onto/into these facilities is restricted, and security is guarded with CCTV monitoring and guard stations. Clearance has to be given by the director of the facility because of homeland security laws and regulations enacted after September 11, 2001.

# Part 8 - Restoration

Mayor Franklin and her staff said the city would need about $750 million to improve city streets, bridges, sidewalks, vehicles, and traffic. City officials proposed using bonds and federal and state grants to help pay for the improvements. The city had spent between fifty and sixty million dollars a year on capital infrastructure projects. Officials said the city needed $262 million[192] for street resurfacing. Fifty-one percent of Atlanta's streets, or 834 miles, are past their life cycle and need repaving. The administration also said the city needed $240 million to repair sidewalks, curbs, and access ramps for the physically impaired. The annual amount the city should budget to whittle away at all its infrastructure needs equals the entire budget of the police department.

In 2009, Mayor Franklin stated, "They better hurry"; she warned that the city's infrastructure backlog hovered around $750 million. Projects are going to pile up and only become more expensive.

Sidewalks Improvement- The city's sidewalks were in desperate need of repair or replacing all over the city, in every area of town from Bankhead, Buckhead, Joseph Lowery Boulevard., Midtown, MLK Jr. Drive, Northside Drive, West End, to West Wieuka. The city had to address broken, cracked, damaged, missing, and uneven sidewalks because of citizen lawsuits and the pressure to comply with the American with Disabilities Act (ADA). Improvements will include the installation of new sidewalks,

---

192    http://m.clatl.com/freshloaf/archives/2011/02/10/fixing-atantas-broken-infrastructure-will-require-patience

ADA-compliant ramps for the handicapped, driveway aprons, new curbing, and the construction of retaining walls.

When construction activities are completed, the contractor crews are responsible for restoring areas disturbed by construction related to their operations. These include:

Landscaping- restoring the areas with seed, straw, or, in some cases, sod. This also includes restoring various sizes and species of trees removed from easements, parks, and private property.

Paving- asphalt paving over road cuts or, in some cases, paving from curb to curb.

Striping- striping crosswalks, parking lots, roadways, and sections in school zones.

Curbing- restoring or replacing granite curbing removed as a result of construction activities.

## Complete Streets[193]

This has not been implemented as approved by the City of Atlanta, but it is discussed here because other cities are incorporating "complete streets" into their legislation.

Complete streets are designed and operated to enable safe access for all users. Pedestrians, bicyclists, motorists, and transit riders of all ages and abilities must be able to safely move along and across a complete street. While no complete-streets legislation currently exists at the national level, the principles of complete streets are becoming increasingly more acceptable in transportation planning, and as of October 2009, 110 jurisdictions in the United States have adopted complete-street–friendly

---

193    http://www.atlantaregional.com/transportation/bicycle-pedestrian/complete-streets

policies. In an article titled "Complete Streets Make Complete Sense[194] Ronald Lall states the first step in developing a complete-street program is to establish and adopt a complete-streets policy. There are about 152 jurisdictions across the United States that have adopted policies or a written commitment to a complete-streets initiative. Some of these jurisdictions are Miami, Charlotte, Louisville, Knoxville, the South Carolina Department of Transportation, the Kentucky DOT, the Florida DOT, the North Carolina DOT. Complete-streets[195] legislation is being considerd in Indiana, Missouri, New York, Texas, Vermont, and Washington. In Georgia, Roswell, Decatur, and Alpharetta have adopted complete-streets policies. The Georgia DOT and the City of Atlanta have yet to adopt a policy or initiate the process to adopt such a policy—and that's where the opportunity lies. The article goes on to state that "now is the time" for the City of Atlanta and GDOT to implement a complete-streets policy, because of the opportunities presented by House Bill 277 (transportation funding).

194    http://theporchpress.com/?p=174

195    www.completestreets.org

# Part 9 - What's the Cost?

The Department of Watershed Management, in 2009 reported an annual operating budget of $543 million per year[196], and listed below are projects that have been completed, are ongoing construction within the department.

*Note:*

Not all the projects and costs are listed but only a portion of work completed. This is a representation of some of the work conducted under the Department of Watershed Management.

## Sewer Separation

| | |
|---|---|
| McDaniel Basin | $ 93.3 Million |
| Greensferry Basin | $ 90 Million |
| Stockade Basin | $ 79.6 Million |

## Sanitary Sewer

| | |
|---|---|
| East Point Relief Sewers and South Camp Hapeville Outfall Replacement Sewers | $ 15.5 Million |
| South River SSO Capacity Relief Sewer Replacement Projects | $16 Million |

---

196    www.atlantawatershed.org/pdf/DWM_AR_2009.pdf

## SSES

Group (SG1) phase1 CIPP-B .............................. $5.5 Million

Sewer Group (SG1) phase 1 PB-B ....................... $8.7 Million

Sewer Group (SG1) phase 2 PB-C....................... $6.1 Million

Sewer Group (SG2) phase 2 CIPP-C................... $6.1 Million

Sewer Group (SG2) phase 2 PB-C...................... $14.1 Million

Sewer Group (SG2) phase 2 CIPP-D .................. $2.2 Million

Sewer Group (SG2) phase 2 CIPP-E................... $7.9 Million

Sewer Group (SG2) phase 1 PB-A....................... $7.9 Million

Sewer Group (SG3) phase 3 SSES-C................... $4.2 Million

Sewer Group (SG5-6) SSES-A........................... $14.6 Million

Sewer Group (SG5-6) SSES-B........................... $13.8 Million

Large Diameter Cleaning ................................. $ 3.9 Million

CIPP- cured in place pipe, PB-pipe bursting, SSES-sanitary sewer evaluation survey

## Milling and Paving

Milling and Paving phase 1 ............................... $5.6 Million

Milling and Paving phase 2 ............................... $9.5 Million

## Tunnels

West Area CSO Tunnel .............................. $286-300 Million

South River Tunnel ............................................ $111 Million

Nancy Creek Tunnel............................................ $150 Million

# CSOs

Custer Avenue ....................................................$28 Million

Utoy Creek Water Reclamation .........................$180 Million

South River .......................................................$84 Million

# Clean Water Atlanta

Stream Bank Stabilization Program .......... $840,000 thousand

Historic Old Fourth Ward Park
Stormwater Detention Pond ..............................$ 14 Million

Valve & Hydrant Program ..................................$ 29 Million

Automated Meter Reading (AMR)
Replacing Program ............................................$ 35 Million

Transmission Main Project ................................. $ ?

Bellwood Quarry ..............................................$ 40.2 Million

# Laboratory

Utoy Creek Laboratory and
Administration Building ....................................$ 21 Million

# Water Treatment Plant

Hemphill Water Treatment Facility ...................$ 36 Million

# Tunnel Pump Stations

South River Tunnel Pump Station ....................$ 112 Million

Intrenchment Creek Tunnel Pump Station .........$ 26 Million

# Greenways

Greenway Acquisition Project ...........................$ 25 Million

# Part 10 - Future Projects

The Department of Watershed Management cannot be complacent about accomplishing a more expansive program that includes projects that may have been designed, planned, or initiated under Mayor Franklin's administration but will be implemented, funded, and constructed under the next mayor, Kasim Reed.

## 1) The Five Transmission Mains Project :

Hemphill-Airport Water Transmission Main

Southwest Water Transmission Main

Fairburn Road Water Transmission Main

Koweta Road Pump Station & Transmission Main

Adamsville Pump Station & Water Main Improvements

## 2) The Leak Detection Program

In March of 2011, the deputy commissioner of public works, Dexter White, announced the Leak Detection and Critical Main program. The Critical Main project will target fifty main segments deemed to be in poor condition or too small to support growth around them. Under the project, the city was divided into four quadrants to address problems such as leaks and water-quality issues. Repairs are also targeting areas where low water pressure has contributed to a drop in the fire-insurance rating. With this program, the city is hoping the average home could save ten thousand gallons of water a year.

## 3)  Valve and Hydrant Program

The Valve and Hydrant Program, a three-year project, will ensure that all valves and fire hydrants in the City of Atlanta, the City of Sandy Springs, and unincorporated South Fulton County are fully operational and reliable. There are roughly ninety-five thousand valves and hydrants in Atlanta's [197] water system, and these assets are critical to both public safety and water-system reliability. Under this program, each valve or hydrant will be located, entered into a Global Positioning System (GPS) database for future access by the Bureau of Drinking Water, and maintained/repaired as necessary. Locating and documenting the water-system assets will allow the department to provide high-quality and more efficient service to customers. In the event of a water-main break, DWM crews will be able to quickly locate valves, repair them as necessary, and prevent a prolonged service disruption to the customers.

## 4)  Water Main Replacement

The City of Atlanta's Water Main Replacement Program[198] is designed to replace and rehabilitate aged and deteriorating water mains throughout the city, including the construction of new water mains in the Georgia Tech and Midtown areas.

Many of the pipes in the system were installed in the early 1900s and are small in comparison with modern standards in water mains. The small size and age of the pipes, coupled with corrosion and sediment accumulation over the years, have affected the flow rate and quality of water in some

---

197    http://www.atlantawatershed.org/projects/valve-hydrants-htm

198    http://www.atlantawatershed.org/watrmain/watermain.htm

Atlanta communities. The Water Main Replacement Program is a part of the Clean Water Atlanta infrastructure program to provide clean, safe water to residents and downstream neighbors. Replacing water mains is necessary business. Work was successfully completed in Garden Hills, Buckhead, and the Capital View community in southwest Atlanta, Spring Street in Georgia Tech Square district, Virginia-Highland, and the Midtown area in 2005 and 2006. Other areas included the Greensferry, McDaniel, and Stockade sewer basins.

To view the maps of the Water Main Replacement Program to see if water mains were replaced in your community visit:

http://www.atlantawatershed.org/watrmain/watermain.htm

Map 1 (FY 2005–2011)

Map 2 (FY 2005–2011)

## 5) Small Meter Audit

The City of Atlanta is committed to ensuring customers receive accurate water and sewer bills. The Small Meter Audit's[199] objectives are to correct data-related issues in the billing system and to identify those meter issues, a process that can only be determined via on-site meter investigations. The desired outcome of the audit is to reduce the number of billing-related concerns, such as crossed meter interface units (miu)[200] numbers, mismatched registers, and system-generated estimates. The audit is a system wide review of water-meter components critical to the accuracy of residential water consumption and customer billing.

Meter Replacement Program- the City will replace or retrofit 150,000 meters with automated meter reading capabilities.

---

199    http://www.atlantawatershed.org/smallmeteraudit/index.htm

200    Accurately transmit water usage from the meter to the automated meter reader (amr) handheld device or mobile data collector.

## 6)   Bellwood Quarry

This property was purchased on June 30, 2006, from Vulcan Materials Company. The site was used as a gravel-mining quarry for most of the last century. The city will use this site as a centerpiece of the Atlanta BeltLine.[201] This property, upon completion with the planned Westside Park, will be the largest park in Atlanta. It will be converted into a 300-foot-deep water-storage reservoir that will give the city a thirty-day supply of drinking water, a forty-five-acre lake, and a 351-acre park for the future Westside Park. The quarry sits on forty acres of land and will take 2.4 billion gallons of water to fill. The city plans to bore an eight-thousand-foot tunnel from the quarry to the Hemphill Water Treatment Plant on Seventeenth Street, a project that will take several years to complete, depending on funding. The plan is a part of the BeltLine extension project.

## 7)   The Atlanta BeltLine Project

The first development of the BeltLine began when the Atlanta & West Point Railroad began building a five-mile connecting rail line from its northern terminus at Oakland City to Hulsey Yard on the Georgia Railroad. The surveys were done and initial construction had begun when the courts ordered a halt in May 1899.

The concept again was picked up in 1999 by a Georgia Tech graduate student (Ryan Gravel) who was working on his master's degree thesis and was frustrated by the lack of transportation alternatives in Atlanta.[202] The plan calls for the creation of a series of parks around the city connected by trail and transit along underutilized and abandoned railroads to serve

---

201    http://www.wsbtv.com/news/giant-quarry-is-centerpiece-for-new-park-on-atlanta/

202    http://en.wikipedia.org/wiki/Belt_Line_(Atlanta)

Atlanta's intown neighborhoods. The BeltLine[203] would connect more than forty neighborhoods, cross more than four thousand acres of industrial land, and go by points of interest such as Piedmont Park, the Carter and King Centers, Zoo Atlanta, the Atlanta University Center, and King Plow Arts Center. It would also push shopping dollars into charming Atlanta neighborhoods.

The BeltLine as proposed under Mayor Franklin would incorporate multi-use trails, a neighborhood-serving transit system (streetcars), parks, foot-paths for non-motorized traffic, bicycling, roller-skating, and walking, creating a continuous path encircling the central part of the city, generally following the old railroad right of way to include planning, affordable housing, community outreach, and awareness.

The BeltLine (a total length of twenty-two miles running about three miles on either side of Atlanta's elongated central business district) is bounded by the Lindberg MARTA station, City Hall East, Grant Park, Atlanta Technical College, West End, Maddox Park, the Hemphill Waterworks, and Piedmont Hospital. It connects the four spokes of MARTA rail lines at five stations.

The plan[204] would expand these existing parks: Enota Park from 0.3 to ten acres; Maddox Park from fifty-two to 114 acres; and Ardmore Park from two to eight acres. The plan would also create these new parks: Peachtree Creek Park, at sixty-five acres; Hillside Park, twenty-eight acres; Holtzclaw Park, to two acres; Historic Fourth Ward Park (formerly North Avenue Park), sixty-three acres; Waterworks Park, 351 acres; and Westside Park, 351 acres—roughly twice the size of Piedmont Park.

---

203    Atlanta Journal-Constitution archives: Cathy Woolard, February 14, 2003 http://docs. newsbank.com/s/InfoWeb/aggdocs/NewsBank/0F932A8B49974DAB/0D57227E12704560 ?p_multi=AJBK&s_lang=en-US

204    www.wikipedia.com

The BeltLine has embarked upon a cooperative partnership with MARTA (the local transit) to complete the first phase of a transit environmental impact study for the entire twenty-two- mile transit corridor. It planned for a BeltLine-MARTA relationship in 2009, and a partnership with the PATH Foundation, which has many years of experience building trails in the Atlanta area.

The project is included in the twenty-five-year Mobility 2030 plan by the Atlanta Regional Commission for improving transit from 2005 to 2030.

Monitoring gridlock traffic on Atlanta's roads in 2003, Mayor Franklin stated that "what we are missing is the infrastructure." The BeltLine would fill a huge gap in our inadequate transit inventory.

## 8)  Historic Fourth Ward Park

Part of the BeltLine project, it is one of the first of five new Atlanta BeltLine parks. Coming in at fifteen million dollars less than the original estimate, it will be the central amenity of the larger park. It broke ground on October 15, 2008 and consists of a storm-water detention pond (holds water) that would help solve some of the historic flooding issues in the area.

It brings seventeen new acres of green space to a neighborhood that lacked this kind of space. It will be an energy-efficient park with new spaces for active and passive recreation, including public art, ball fields, and the city's first skating park, which won a grant of twenty-five thousand dollars t from the Tony Hawk Foundation.

Note: To view the proposed BeltLine tour routes visit:

*(http://beltline.org/portals/26/maps/Tour/blp_map_southwest_final.pdf)*

## 9)   Clear Creek Combined Sewer Capacity Relief Project

The City of Atlanta's Department of Watershed Management (DWM), in conjunction with Atlanta BeltLine Inc. (ABI), will construct a storm-water detention pond in the Historic Fourth Ward Park that will serve as a functional amenity for the surrounding community. The city is undertaking the Clear Creek Combined Sewer Capacity Relief Project as part of a federal Consent Decree. The storm-water pond will be integrated into the initial phase of the Historic Fourth Ward Park (H4WP). In September of 2007, the Atlanta City Council approved the transfer of thirty million dollars from the DWM to ABI to allow it to design, manage, and construct this phase of the project on behalf of the city. ABI will design and fund future phases of the H4WP project.

The project includes construction of a nine-million-gallon storm-water detention pond that will capture storm-water runoff from an approximately 800-acre drainage area upstream of the North Avenue location. Once completed, the detention pond will provide peak-flow attenuation to the Highland Combined Sewer Trunk and capacity relief to the overall Clear Creek Combined Sewer Basin. The original design of the project called for the construction of an underground conveyance tunnel at an estimated forty million dollars. However, the creation of the storm-water pond creates a functional, creative, and sustainable feature for the neighborhood and a cost savings for the Department of Watershed Management.

## 10)  South River Relief Sewer- North Basin

Four sewer lines totaling 11,200 linear feet will be upsized to alleviate the current hydraulically overloaded condition and to accommodate future sanitary flow requirements.

- Lower McDaniel and Outfall Replacement- 6,150 linear feet of existing 24"–30" will be upsized to 36"- 42" diameter piping to relieve system overloading.

- Lower Tenth Ward Relief Sewer- 2,700 linear feet of existing 36" diameter piping will parallel with a second 36" diameter pipe.

- Lower Tenth Ward Trunk Replacement- 1,800 linear feet of existing 36" diameter piping will be upsized to 42"- 48" to relieve system overloading.

- Arthur Langford Jr. Place SW Outfall Replacement- 410 linear feet of existing 8" diameter piping will be upsized to 10" diameter piping to relieve system overloading.

## 11) South River Relief Sewer- South Basin

Two sewer lines totaling 9,200 linear feet will be upsized to alleviate the current hydraulically overloaded condition and to accommodate future sanitary flow requirements.

- Jonesboro Trunk- 4,800 linear feet of existing 30" will be upsized to 36" diameter piping to relieve system overloading.

- Forest Park Outfall- 4,400 linear feet of existing 12"- 15" will be upsized to 18"- 24" diameter piping to relieve system overloading

## 12) South River Basin- East Point Sewer Replacement

The work is limited to the East Point Trunk, South Camp Hapeville, and South River Outfall Replacement Sewers. The project limits include fifteen thousand linear feet of pipeline, of which 1,900 linear feet will be installed using trenchless technologies. The entire sewer line will be upsized to alleviate the current hydraulically overloaded condition to accommodate future sanitary flow requirements.

- East Point Trunk- approximately 10,000 linear feet long. The existing diameter pipe will be upsized to 42" diameter piping to relieve system overloading. The upper extension to this main trunk will be approximately 1,079 linear feet long. Existing diameter pipe will be upsized to 36" diameter pipe to relieve system overloading.

- South River Outfall- approximately 1,120 linear feet long. The existing 10" diameter pipe will be upsized to 24" diameter pipe.

- South Camp Hapeville Outfall- approximately 1,645 linear feet long. The existing 10" and 12" diameter pipe will be upsized to 18" diameter pipe.

## 13) Sale of City Hall East

In January 2012, Mayor Reed and the City of Atlanta completed the sale of City Hall East to Jamestown Properties.[205] The old historic Sears and Roebuck Company building site will be redeveloped and called the Ponce City project, which will be the largest adaptive reuse project in Atlanta's history.

Proposals were sent to the city initially in March 2004 and finally in 2012 the city was able to seal the deal to Jamestown Properties through the Atlanta Development Authority for twenty-seven million dollars, including an initial $15.5 million payment at closing. Mayor Reed states "I am pleased we have completed the transaction on City Hall East. This building has been one of Atlanta's signature landmarks for more than eight decades, and I am confident that this historic building is in good hands with this world-class firm. I look forward to seeing the City Hall East transform into

---

205    www.southeasterngreen.com/index.php/metro-atlanta-news/4367-mayor-kasim-reed-and-the-city-of-atlanta-complete-sale-of-city-hall-east-to-jamestown-properties

a top destination for residents and visitors and become the heart of the resurging Ponce de Leon corridor."

## 14) Collier Road Water Main Extension Project

Starting in July 2012, the City of Atlanta was expected to begin work to install a new water main on Collier Road. The project was intended to improve the water supply to Piedmont Hospital. A month long project, the work included:

- Connecting the 16" water main on the West side of the hospital and a new 16" water main on Collier Road at Dellwood Drive, and connecting the 16" main at Arden Drive.

## 15) City of Atlanta 311 Center

The City of Atlanta was awarded $3.3 million from Bloomberg Philanthropies[206] to fund an Innovation Delivery Team to generate innovative solutions, develop implementation plans, and manage progress towards defined targets. This amount, down from the previous commitment of $4.2 million from New York Mayor Michael Bloomberg[207] through his Bloomberg Philanthropies to be used to address civic needs, such as a the new 311 call center.

311 call centers have been successfully used in other cities to provide citizens one call center through which many city services can be reached. The purpose is that because 911 call centers are disrupted by nonemergency calls, which undermine their effectiveness, 311 call centers would act to better inform citizens while reducing call volumes that interrupt operations in the 911 call centers.

---

206    http://targetednews.com/nl_disp.php?nl_date_id=241659

207    http://iq.callme.io/2011/07/21/atlanta-to-receive-4-2-million-from-bloomberg-for-311-call-center/#

# Part 11 - Accomplishments

Department of Watershed Management

The City of Atlanta has developed a fifty-year Master Plan for Water Resource Management, developed and implemented a long-term comprehensive financial plan, invested two billion dollars in water and wastewater capital improvements, and created a valve/hydrant program to evaluate if valves and hydrants are functional and when they need replacing. The department has replaced more than one hundred miles of water mains and is performing a comprehensive assessment and rehabilitation program to improve system performance and reduce water loss. The department has cleaned more than 2.4 linear miles of sewer pipe. And it has turned into a utility business with annual budgets and a four-billion-dollar capital program.

The 2009 annual report listed the accomplishments of people, programs, and offices within the Department of Watershed Management that have set the benchmark in achievement for clean water and infrastructure work for the City of Atlanta.

## Sewer and Water Awards:

GWPCA 2002 Gold Award- R.M. Clayton, Utoy Creek, and South River Water Reclamation Centers. These facilities operated during the calendar year with no permit violations.

NACWA for wastewater-treatment plants: Hemphill/ Chattahoochee.

Platinum Award- Five years. Compliance in state and federal standards.

AWWA for drinking water.

NACW for drinking-water treatment plants- Gold Award.

AMWA Platinum award for utility excellence (attributes of effectively managed utilities).

AMWA from the Chattahoochee Riverkeeper.

AMWA for Utoy Creek wastewater facility (state-of-the-Art water-quality testing lab).

Certified Accreditation Program- Utoy Creek waste-water lab ( EPA)

WEF- Laboratory Analyst Award 2007 to Robert Williams, assistant laboratory manager, for enhancing the global water environment.

AMWA- for watershed management.

GAWP- Gold Award for operational excellence.

NACWA Platinum Peak Performance Award for Utoy Creek and South River Reclamation Centers.

Platinum Peak Performance Award- 10 years. Compliance with federal water pollution standards.

GAWP- Ira C. Kelly Award for outstanding service in laboratory operations.

COE- CDR for "Customer Call Center of Excellence" 2008, through service delivery and system reliability.

AWWA Wastewater Reclamation Center Awards- DWM earned three Platinum, seventeen Gold, and three Silver national awards.

> ➢ NACWA- National Association of Clean Water Agencies
>
> ➢ AWWA- American Water Works Association
>
> ➢ NACW- National Association of Clean Water
>
> ➢ AMWA- Association of Metropolitan Water Agencies
>
> ➢ WEF- Water Environmental Federation

➤ GAWP- Georgia Association of Water Professionals

➤ GWPCA- Georgia Water and Pollution Control Association

## Other City Awards include:

Marvin M. Black 2002, Excellence in Partnering Award

Phoenix Park II- Park of the Year 2006, City of Atlanta

Freedom Parkway designated Atlanta Public Art Park 2007

Greenway Protection Award- 2007, City of Atlanta (Greenway Acquisition Project)

National Award for Entrepreneurial Leadership 2007, City of Atlanta

Atlanta named Most Playful City 2007, KaBOOM!

Water Efficiency Leader Award- 2008 Robert Hunter

# Part 12 - Sustainability

In 2006, Mayor Franklin identified environmental sustainability as a critical factor in making Atlanta a "best in class" city.[208] In early 2007, she charged her administration to create and foster a community dedicated to sustainability through best-in-class leadership and to implement solutions and practices as they are identified. To answer the charge, a team of city officials and expert consultants had assessed Atlanta's current sustainable practices and made recommendations for a course of action in 2008 for new high-impact programs and policies.

## Sustainable Atlanta[209]

Sustainable Atlanta was founded in 2007 out of the city's commitment from Mayor Franklin for long-term economic vitality and environmental prosperity. For the first time, diverse groups of stakeholders are coming together to address the sustainability concerns that Atlanta faces. Leaders and experts from the environmental, business, and government field convene regularly to develop policies and strategies that will deliver lasting solutions for our community. Leveraging the national, state, and metro Atlanta regulatory framework as springboards to future progress, we will go above and beyond what is required to make Atlanta truly sustainable.

---

208    City of Atlanta online; http://174.37.215.145/mayor/sustainableatlanta.aspx

209    www.sustainableatlanta.org

Environmental sustainability is a critical factor in making Atlanta a more competitive and viable city. Sustainable Atlanta's website states that a vision for a sustainable Atlanta would include that:

- The air is clean.

- Waste is reduced.

- Residents use less water than what is available.

- All buildings are high-performance structures.

- Green space is abundant.

- Green industries and businesses thrive.

- Clean energy technologies prevail.

- Solutions are equitable and inclusive.

- Transportation alternatives are the norm.

- Communities and urban centers are walkable.

- Quality of life is ensured for future generations.

In a 2008–2009 sustainability report for Atlanta, then executive director Lynnette Young, stated that the report for Atlanta is both a map and a milepost. "It measures our initial steps so that future decisions will be more strategic and future actions more efficient." The areas of focus are:

## Water-

Sustainable Atlanta is focused on a fifty-year vision for water, concentrating on key steps within the next three years. A comprehensive approach is necessary, one focused both on supply (infrastructure) and demand (use and efficiency).

## Energy and Climate Change-

Sustainable Atlanta's initial focus on energy and climate change falls into three distinct areas: sustainable buildings and development; air quality and transportation by creating the Atlanta Sustainable Building Task Force to set fossil-fuel reduction targets for all new buildings and major renovations; and reducing use incrementally until achieving carbon neutrality in 2030.

## Parks and Green Space-

In May of 2004, the Department of Parks, Recreation and Cultural Affairs (DPRCA) embarked on a strategic planning[210] process to embrace the living values under Mayor Franklin's administration and create a mission and vision for the department.

A five-year strategic plan of action was implemented for overall service improvements and to bring the DPRCA close to the mayor's goal of 'Best in Class" for the city.

The City of Atlanta has 338 parks (see diagram) spread over between 3,400 to 3,570 acres throughout the city, approximately 3.8 percent of the city's land area.

The Bureau of Parks maintains 248 of these parks; the remainder is either for adoption or has already been adopted by volunteer organizations. There are 108 playgrounds located in 104 of the park areas, twenty pavilions, six golf courses, thirty recreation facilities, twenty swimming pools (including five natatoriums). Among the nation's twenty-five largest cities, Atlanta has the least amount of land dedicated to parks. Atlanta currently provides 7.5 acres of parkland per one thousand residents and protects at least 75 percent of environmentally sensitive lands through ownership and/or development regulations.

---

210    http://www.atlanta.gov/modules/Showdocument.aspx?documentid=910

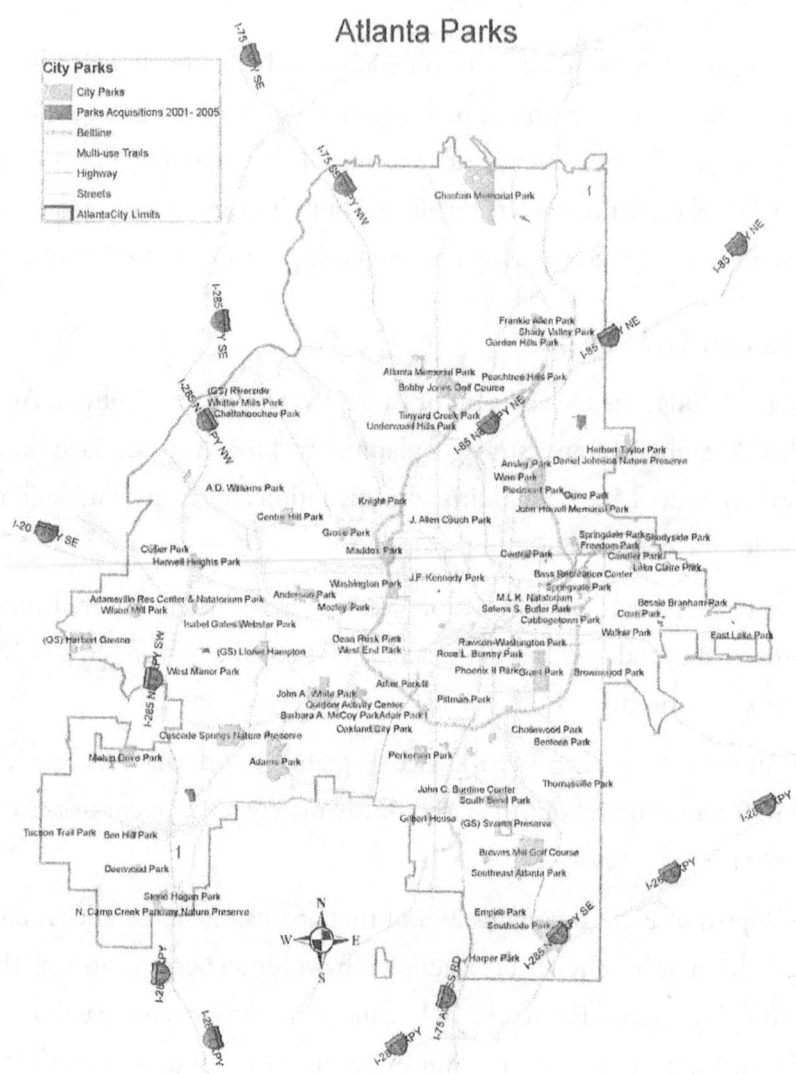

Atlanta Parks

# Part 13 - Going Green

## The Greenway Acquisition

Georgia State Bill 399 was passed and formed a Greenspace Commission to protect communities' green space. Its adoption enables communities to preserve at least 20 percent of their land areas as connected and open green space to be used for informal recreation, such as picnics, camping, and hiking trails, and natural resource protection. The departments chosen to oversee the various projects around the city are the Department of Planning and Community Development and the Department of Parks, Recreation, and Cultural Affairs.

Another component to the bill is the Quality of Life Program, which allocates money for projects around the city, including public plazas, walkways, bike trails, and park amenities to allow for public access to more green-space activity areas such as Greenway corridors (linear parks) and connect to all major parks, stream corridors, and public spaces.

As part of the Consent Decree, the city is to set aside space for "Green Space." The city purchased 38.5 acres out of a total of two thousand in southwest Atlanta, topping its 460-acre mark in its greenway acquisition program. The area, which is partly a flood plain, is intended to stem downstream flooding by slowing the flow of storm water into the creek. Under the twenty-five-million-dollar project, property in Atlanta and fourteen other counties in the metropolitan area are being acquired and twenty-three miles of stream bank protected till 2007.

The intent of the Greenway Acquisition Project was to acquire stream-side buffer areas to protect streams and creeks from erosion from encroaching development. The buffer areas provide a vegetation filter using the natural growth of vegetation near streams to protect the streams from runoff so the designated streams may be maintained or restored to their natural condition. The plan also protects animal habitats, wetlands, and an ecosystem for fish.

A greenway established adjacent to a river or lake provides a filter through which surface water runoff passes before reaching the receiving river or lake. As surface water runoff passes through the greenway, its velocity is reduced due to resistance of the ground's natural vegetation.

In September 2002, the city hosted a Greenspace Cleanup of Swann Preserve, which the city purchased on a fifty-seven-acre site to be used as a nature preserve under the Greenway Acquisition.

According to the Consent Decree, failure of the city to perform any obligation or to observe or fulfill any condition of the Greenway Acquisition Plan shall be deemed a failure to comply with the requirements of the Consent Decree.

The program is to acquire and protect properties adjacent to selected rivers and creeks within the Metro Atlanta area. Once acquired, these properties (Greenway Properties) will forever be maintained in a natural, undisturbed state.

The conservation of Greenway Properties will protect water quality in rivers and streams and will also protect animal habitats, plant habitats, and wetlands along Metro Atlanta Rivers and streams. In addition, land adjacent to the waterways will be protected from erosion, flood damage, and clear-cutting.

The City of Atlanta's Greenway Acquisition Plan[211] encompasses the document as spelled out in accordance with the requirements of Section VIII of the Consent Decree of 1998.

The Greenway Planning Process consists of three phases:

1) Inventory Phase

2) Assessment Phase

3) Planning Phase

The Greenway system[212] encompasses stream segments of the Chattahoochee River Drain Basin along Nancy Creek, Proctor Creek, Clear Creek, Utoy Creek, South Utoy Creek, Camp Creek, Bear Creek (East), Sweetwater Creek, Anneewakee Creek, Dog River, North Dog River, Wolf Creek, and Snake Creek. It also includes South River Drainage Basin at the South River, Intrenchment Creek, and Honey Creek.

The city will continue to purchase parcels throughout the city and the fourteen surrounding counties, including Butts, Carroll, Clayton, Cobb, Coweta, Dekalb, Douglas, Fulton, Gwinnett, Heard, Henry, Newton, Paulding, and Rockdale. The parcels will be used to create buffer zones to protect stream banks and stop urban development that encroaches down to the creeks and streams. Once the target areas were identified, the city encouraged property owners to either donate or sell to the city a conservation easement, which is a voluntary, legal agreement between the property owner and the City of Atlanta.

---

211    http://www.cleanwateratlanta.org/greenway/GreenwayPlan/default_main_main.htm

212    Greenway Acquisition Plan- US Infrastructure, Feb 2001

## Project Greenspace[213]

The City of Atlanta has an initiative whose vision is a long-term plan for growing and managing Atlanta's green-space system to create a vision and framework for a world-class system that connects people with great public spaces, nature preserves, parks, plazas, and streetscapes.

In December 2004, Atlanta adopted the New Century Economic Development Plan. One of the plan's initiatives focused on parks and green space.

Leadership: Sustain the Movement-

"It's all about changing the way we behave," as its website states. It's about choosing to preserve more and throw away less. It's about finding healthier ways to commute. It's about designing and building better so that we live and work in more efficient and healthier buildings. It's about changing what you do and convincing your colleagues and neighbors to do the same.

How SustainLane Ranks Atlanta (2008–2009)

SustainLane, an online guide to sustainability and just one form of independent measurement, ranked the Atlanta area as the No.19 greenest city among the nation's top fifty most populous cities. The report continued to state that Atlanta led the Southeast in environmentally friendly buildings. Atlanta's SustainLane ranking (out of 50 states):

**Energy and Climate Change**

Green Building/LEED- No.3 (sustainability leader)

Metro Transit Ridership- No.11 (sustainability advance)

City Community- No.17 (sustainability advance)

Energy and Climate Change Policy- No.18 (sustainability advance)

---

213    http://www.atlantagreenspace.com

Air Quality- No.42 (sustainability in danger)

Metro Street Congestion- No.45 (sustainability in danger)

**Water**

Water Supply- No.17 (sustainability advance)

Water Quality-No.40 (sustainability challenged)

**Recycling**

Solid Waste Diversion- No.29 (mixed)

**Parks**

Planning and Land Use- No.42 (sustainability in danger)

From the sustainability report for Atlanta, which emphasizes lead, change, and sustain, the essential question is: "How does an area as big and vibrant as Atlanta retrofit itself so that it can accommodate growth with its existing systems, road, buildings, culture and infrastructure"?

The population in and around Atlanta has swelled beyond five million people. Growth has brought about its challenges. For the people who work and play in Atlanta, the area is home to the nation's third-longest commute time. The daily grind on the roadway symbolizes our deteriorating air quality and our increasing urban sprawl. The report goes on to say sustainability is not a new concept, but there is renewed energy and commitment to the movement because of changes in our world and our increased understanding of our impact on our region. Mayor Franklin's vision is one of Atlanta as a recognized leader on sustainability.

## Partnerships with Organizations

The City of Atlanta will also seek partnerships with organizations, agencies, and corporations that have complementing goals and are willing to share ideas and consider joint-venture projects to accomplish such goals for the environment and communities.

### ❖ The PATH Foundation

Out of an idea in 1991, the Path Foundation[214] has created a network of off-road trails in and around Atlanta for walkers, runners, cyclists, and skaters, and a series of scenic greenways to preserve our region's forested character and offer opportunities for families to enjoy nature together. The trails and greenways were designed as a way to connect neighborhoods to each other, to get people out of cars, to encourage healthier lifestyles, and to improve our quality of life.

PATH trails enhance community spirit and bring neighborhoods together. Each day, thousands of joggers, walkers, bikers, and skaters from all walks of life escape the roads and hit the trails for travel and recreation.

PATH has made significant progress toward building Georgia a network of trails, including: Silver Comet, Stone Mountain, Lionel Hampton, Westside, Arabia Mountain, Chastain Park, South Decatur Trolley, Northwest Atlanta, and Freedom Park Trails.

### ❖ Atlanta Strategic Action Plan (ASAP)[215]

This group formed in 2005 under the name DCA, and full adoption came in the fall of 2009. In the group's 2007 presentation to the city, "Why we are here" it highlighted its function and purpose—Atlanta becoming a world-class city with the focus on population, economic development, housing, transportation, community facilities, natural/cultural resources, land use, and involvement in neighborhood planning units.

---

214  http://pathfoundation.org/about

215  http://www.atlantaga.gov/government/planning/asap.aspx

## ❖ Neighborhood Planning Unit (NPU)

The Neighborhood Planning Unit is a neighborhood–scale government structure established in 1974 by the city's first African American mayor, Maynard Jackson. His aim was to ensure that citizens, particularly those who had been historically disenfranchised, would be in a position to comment on the structure of their communities, and to ensure that the citizens would not have this ability stripped of them by politicians who found an involved and engaged public inconvenient.

The City of Atlanta is divided into twenty-five neighborhood planning units lettered from A to Z, except U; each unit represents the citizens in a specified geographic area, and it is an advisory council that makes recommendations to the mayor and the city council on zoning, land use, and other planning issues. The NPU system provides an opportunity for citizens to participate actively in the Comprehensive Development Plan, which is the city's vision for the next five, ten, and fifteen years. It is also used as a way for citizens to receive information concerning all functions of government. The system enables citizens to express ideas and comments on city plans and proposals while assisting the city in developing plans that best meet the needs of their communities. Each NPU meets once a month to review applications for rezoning properties, varying existing zoning ordinances for certain properties, applications for liquor licenses, applications for festivals and parades, any changes to fees charged by the city, any changes to the city's Comprehensive Development Plan, and any amendments to the City's Zoning Ordinances. Once an NPU has voted on an item that vote is then submitted to the relevant body, which makes the ultimate determination with regard to that issue as the official view of the community on a topic. NPUs operate according to a varied set of guidelines. Each NPU is permitted to create its own bylaws. NPUs operate in a representative governmental fashion, with only elected representatives voting on the issues at hand. Each NPU is assigned a City of Atlanta planner who attends the monthly meetings. Planners are charged with recording official votes, responding to questions about issues of land use and zoning. The NPUs are staffed entirely by citizen volunteers, who receive no compensation for their efforts.

## ❖ Atlanta Pedestrian Safety Group

PEDS[216] is a results-oriented pedestrian advocacy organization making metro Atlanta pedestrian-safe, friendly, and accessible for all pedestrians.

Founded in 1996 by Sally Flocks, her efforts have prompted changes at local, regional, and state levels in Georgia and have made PEDS one of the most effective transportation reform organizations in the United States. She and her staff work with government officials, community leaders, and the media to overcome barriers to walking, including organizing volunteers, assisting with outreach, and gathering information on best practices for bus-stop locations and safe crossings at transit stops.

Their goals are:

- Change community attitudes to favor pedestrians.
- Increase walking and other pedestrian activities.
- Ensure the design of pedestrian-oriented communities.
- Advance the equitable use of transportation funds.
- Make sidewalks and crossings safe and accessible to all users.
- Reduce the risk of injury and death for pedestrians.

*Note:* Also see Pedestrian Advocacy Group- Where the Sidewalk Starts."[217]

---

216    http://peds.org/about_peds_achievements

217    http://www.wherethesidewalkstarts.com/p/pedestrian-advocacy.html

This is a pedestrian advocacy group that works to address the problems in many urban cities that need or have overlooked pedestrian safety in transportation planning.

## ❖ CAP/ADID

Central Atlanta Progress Inc. (known as CAP), is a private nonprofit community development organization providing leadership, programs and services to preserve and strengthen the vitality of downtown Atlanta. It works side by side[218] with the Atlanta Downtown Improvement District (known as ADID), a public private partnership that strives to create a livable environment for downtown Atlanta, in alliance with Mayor Franklin's vision for a thriving downtown district. This district contains 220 blocks within an area generally bound by North Avenue on the north, Memorial Drive on the south, Piedmont Avenue, and the Downtown Connector on the East, and the Norfolk-Southern rail line on the west. For the years 2002–2010, the collaboration was instrumental in creating a banner program to enliven downtown streets; a Downtown Development Day to showcase new housing, commercial and entertainment developments, and quality office space; "Let's Do Downtown," a one-hundred-day downtown initiative launched in partnership with the city; and "Imagine Downtown," an eight-month strategy-planning process for downtown. Downtown Atlanta's In-Bloom program installed and set up maintenance for flower baskets throughout downtown. CAP/ADID was integral in working with Mayor Franklin and Governor Perdue in the  construction of the Ivan Allen Jr. Boulevard to improve east-west access in downtown. It also implemented the downtown and midtown Wayfinding Signage System that installed over 270 signs.

In 2009, as part of the first phase of the Master Plan implementation, ADID did hardscape improvements to repair Woodruff Park by replacing

---

218    http://www.atlantadowntown.com/about

missing and broken pavers, and cracked and missing seat walls through-out the park. Also added was the recreation and reading room area. This structure holds equipment for games, books, magazines, and supplies for special events.

## ❖ TEAM HOME DEPOT

In September of 2009, Kelly Caffarelli, president of the Home Depot Foundation,[219] Park Pride, and volunteers rolled up their sleeves to help Mayor Franklin to clean up, refurbish, and revitalize Spink-Collins Park. The goal was to transform the park, whih was previously an overgrown, forgotten, 21.49-acre property, it into a signature green space in Atlanta. The spruce up included laying sod, building a series of looping soft-and-hard surface paths, boardwalks, an outdoor classroom, a grass meadow, and a small pavilion to anchor the trails. The Home Depot signed on as the largest partner of Greener Atlanta Initiative.

---

219    http://www.homedepotfoundation.org/celebration_highlights.html

# Part 14 - Green Infrastructure

The City of Atlanta has not completed the cycle on "Going Green" as of yet without totally incorporating green infrastructure. Green infrastructure refers to storm-water management. The next step for Atlanta is to create green infrastructure zones (or corridors) throughout the city.

## What is Green Infrastructure?

Green infrastructure helps stop runoff pollution by capturing rainwater and either storing it for use or letting it filter back into the ground, replenishing vegetation and groundwater supplies. Examples of green infrastructure include green roofs (vegetated/eco-roof), street trees, increased green space, rain gardens, and "permeable pavement."[220]

Green infrastructure is strategically planned and managed networks of natural lands, working landscapes, and other open spaces that conserve ecosystems' values and functions and provide associated benefits to human populations. Green infrastructure provides a solution that ensures environmental protection and a higher quality of life within communities.

Critical elements of the implementation strategy, such as low-impact development practices (LID), conservation, developments, and green/grey

---

220  A range of materials for paving roads, bicycle paths, parking lots, and sidewalks that allow the movement of water and air around the paving material, whether concrete, asphalt, paving stones, or brick.

interface, are necessary components to any successful green infrastructure plan.[221]

These solutions have the added benefits of beautifying neighborhoods, cooling and cleansing the air, reducing asthma and heat-related illnesses, lowering heating and cooling energy costs, boosting economies, and supporting American jobs.

It is a new term but not a new idea. The concept originated in the United States in the mid 1990s and highlights the natural environment in decisions about land-use planning. The US Environmental Protection Agency (EPA) has extended the concept to apply to the management of storm-water runoff at the local level through the use of natural systems or engineered systems. Green infrastructure refers to systems and practices that use or mimic natural processes to infiltrate, evapotranspire,[222] or reuse stormwater runoff on the site where it is generated (Green Infrastructure Action Strategy, American Rivers et al., 2008). It is the interconnected network of natural areas and other open spaces that conserve natural ecosystems and function and provide associated benefits to human populations. (Green Infrastructure, Benedict & McMahon, 2006). It is our nation's natural life-sustaining system. Green infrastructure[223] differs from conventional approaches to open-space planning because it looks at conservation values and actions in concert with land development, growth management, and built-infrastructure planning.

City planners and neighborhood planning units (NPUs) in Atlanta will have to get together to walk their districts to come up with creative ideas on where to implement storm-water management strategies to create

---

221    http://www.greeninfrastructure.net/content/definition-green-infrastructure

222    A term used to describe the sum of evaporation and plant transpiration from the earth's land surface to the atmosphere.

223    www.sprawlwatch.org/greeninfrastructure.pdf

curbside swales and curb extenders that capture storm water on site rather than sending it into storm drains; to create tax rebates for residents who install rain barrels on and around their property; and to create green rooftops (vegetated/eco-roofs) in the downtown business district, Buckhead, Midtown, and Virginia-Highland, and around the Georgia Dome and Turner Field areas. They need to incorporate in zoos, parks, parking lots, driveways, and around museums garden paths (hardscapes) made from paving stones that allow rain water to infiltrate into the ground. They need to install storm-water planters that collect storm-water runoff from roofs, streets, driveways, and sidewalks that slow the flow and allow water to soak into the ground as soil and vegetation filter pollutants. Buildings could collect and reuse storm water to replace potable water for use in flush fixtures, helps in managing storm water and reducing water use. The use of drip irrigation, adaptive plantings, and replacing perennials with native plants is a strategy for water efficiency. Native plants can be placed in curbside swales because they require less fertilizer and less pesticide.

For example, curbside swales[224] could be installed off heavily sloped roads that dump storm water out into the roadway causing standing water in places like International Boulevard, Sidney Marcus Boulevard, Joseph Lowery Boulevard, Northside Drive, MLK Jr. Drive, West Peachtree Road, and areas along Peachtree Road.

By diverting storm water from drains and runoff, the storm water can be re-used for natural irrigation for plants and wildlife. Green infrastructure is the interconnected network of open spaces and natural areas—greenways, wetlands, parks, forest preserves, and native vegetation—that naturally manage storm water, reduce the risk of floods, capture pollution, and improve water quality. Planting trees will provide cover for hardscapes

---

224   Natural hydrological systems that capture storm water on site rather than sending it into pipes.

and can also reduce storm-water runoff. Ground-up brick can be used for drainage as a reusable material. Clustering together commercial and residential developments helps improve the quantity of storm-water runoff. Arainwater[225] harvesting system and rain gardens can help with water efficiency and storm-water management, as does the use of open grid[226] paving that contains vegetation in the open cells. Paint downtown office-building roofs and parking lots with a high SRI coating[227] to reduce the heat-island effect.[228] Planting with native plants, micro misters, and adaptive plants is called water-efficient landscaping. The collection of wastewater from sinks (not toilets), laundry machines, or showers (gray water) can sometimes be recycled for irrigation, depending on the local codes. The city should be adding wildlife refuges which are the best storm-water receptacles a city can have.

The city could give awards to businesses and citizens that come up with creative green ideas that benefit the community, address storm-water containment, runoff, and sediment control, or use landscaping strategies that best eliminate the need for irrigation using xeriscaping.[229]

The city should force contractors working on large construction sites to build retention ponds at the beginning of a notice to proceed to contain erosion and sediment once the site has been cleared, thus eliminating further washout into the roadway.

Cities are currently using green infrastructure to control storm-water management practices. More recently, studies are now being done to see how vacant properties can be converted by "Right-Sizing through Land Banking

---

225   Non potable (cannot drink).

226   Pavement that is less than 50% impervious.

227   A material with an SRI of 100 is light colored.

228   Reduces the amount of area that is dark.

229   A method that employs drought-resistant plants or eliminate water use.

and Green Infrastructure." This right-sizing model relies on a menu of green-infrastructure strategies in conjunction with the innovations in the control and management of vacant properties through land banking and even urban land trusts.

Roof Tops to Rivers[230]and Roof Top to Rivers II, profiles "14 cities"[231] of all sizes using "green infrastructure"—a set of design strategies that mimic nature's own hydrology and allow rain to filter back into the ground right where it falls, to tackle storm-water pollution and sewage overflows. The cities profiled are using vegetation around parking lots, rain gardens, green roofs, permeable pavement, and trees to help absorb the water like sponges. By using green space, swales, cisterns, and other techniques, green-infra-structure solutions bestow a range of benefits on communities that embrace them. The article goes on to state that a 2007 EPA study found that in the vast majority of cases green-infrastructure practices save money for developers, property owners, and communities while protecting and restoring water quality.

An estimated ten trillion gallons a year of untreated storm water runs off roofs, roads, parking lots, and other paved surfaces, often through the sewage systems, into rivers and waterways that serve as drinking supplies, and flows onto our beaches, increasing health risks, degrading ecosystems, and damaging tourist economies.[232] Cities of all sizes are saving money by employing green infrastructure as part of their solutions to storm-water pollution and sewage overflow problems.

---

230    http://switchboard.nrdc.org/blogs/dbeckman/report_cities_nationwide_using.html

231    Aurora, Bellevue, Chicago, Kansas City, Milwaukee, Nashville, New York, Orlando, Philadelphia, Pittsburg, Portland, Rouge River Watershed, Seattle, Syracuse, Toronto, Washington, DC.

232    http"www.nrdc.org/water/pollution/storm/chap7.asp

## How does Green Infrastructure Benefit the Environment?[233]

Green infrastructure is associated with a variety of environmental, economic, and human-health benefits, many of which go hand in hand with one another. The benefits of green infrastructure are particularly accentuated in urban and suburban areas, where green space is limited and environmental damage is more extensive. Green-infrastructure benefits include:

- Reduced and delayed storm-water runoff volumes

- Enhanced groundwater recharge

- Storm-water pollutant reductions

- Reduced sewer overflow events

- Increased carbon sequestration

- Urban heat-island mitigation and reduced energy demands

- Improved air quality

- Additional wildlife habitat and recreational spaces

- Improved human health

- Increased land values

*Notes:* It has been determined that landscape irrigation consumes 30 percent of potable water[234] in the United States each day on average.

## Case Studies:

These cities and states have already gotten started.

1) Portland, Oregon- The Portland Green Street Program.[235]

---

233    http://cfpub.epa.gov/npdes/home.cfm?program_id=298

234    Water that is safe to drink.

235    http://www.portlandonline.com/BES/index.cfm?c=44407

2) Kansa City, Missouri- MetroGreen[236]

3) Baltimore, Maryland- The Baltimore County Forest Sustainability Project

4) Florida- Florida's Ecological

5) Minnesota- The Metro Greenways: Seven-County Twin Cities Region

6) Saginaw Bay, Michigan- The Saginaw Bay Greenways Collaborative

7) Northwest Lower Michigan- The Conservation Resource Alliance's Wild Link and River Care Programs

8) Sonoran Desert, Arizona- The Sonoran Desert Conservation Plan Pima County

9) Seattle, Washington- The Mountains to Sound Greenway

10) Massachusetts- The Massachusetts BioMap Project

11) Chicago- Green Alley Program[237]

12) New York- Flushing and Gowanus Green Infrastructure Grant Initiative[238]

Outside the United States, according to the Infrastructure 2010 report,[239] Australia is the model for water conservation, storm-water capture, and recycling, as well as more condensed land-development practices, using a combination of basic and sophisticated techniques that could be applied in US cities and others globally.

---

236    http://greeninfrastructure.net/gi_case_studies

237    http://tlc.howstuffworks.com/family/dep-26-million-stormwater-capture-projects-nyc.htm

238    http://www.nyc.gov/html/dep/html/news/dep_stories_p1-29.shtml

239    http://www.sustainablebusiness.com/index.cfm/go/news.display/id/20145

# Part 15 - What's Next?

## Traffic and Transportation

In Atlanta, congestion is getting worse. People are moving closer to the center core of the city, with new condominiums and apartments popping up, new work-and-play communities like Atlantic Station. In addition, business, banking centers, and office complexes are building around transportation stations like our MARTA parking decks, increasing the density of the urban core and placing new demands on the transportation infrastructure.

The goal of the city is to establish integral multimodal transportation systems that move people and goods in an efficient and environmentally sensitive manner[240].

As stated in this article, the city's transportation policies focus on improving and enhancing mobility in a manner that improves air quality and creates alternative travel, such as foot, bike, transit and cars.

Atlanta, the state of Georgia (GDOT), and the regional agencies ARC, GRTA, GRPA, and DCA have come up with creative strategies to address the equity in rebalancing transportation and land-use policies and funding to improve and enhance the travel needs of the region, including the linear Core, the Arterial Core, Transit, Citywide Development and Transportation Initiatives, and Livable Center Initiatives.

---

240    City of Atlanta 2004 CDP: Transportation

In the transportation sector, airplanes and automobiles continue to have the biggest impact on the metropolitan region.[241]

Automobiles and the transportation modes that had done so much to shape the city in the first half of the twentieth century continued to affect the layout and lifestyles of metropolitan Atlanta in the last decades of the century. As the number of residents in the outlying areas continues to grow, the Georgia Department of Transportation responded by increasing the number of passenger lanes on Interstates 75, 85, and 20 (including, most recently, the construction of high-occupancy vehicle (HOV) lanes to encourage carpooling) and also on Interstate 285, the high-speed, limited-access highway that encircles the city. Georgia 400—a toll road connecting suburban communities north of Atlanta to the city—was also constructed. By the turn of the century, 2.5 million vehicles were registered in the metropolitan area, and motorists were driving approximately one hundred million miles every day on Atlanta highways and roads.

Officials for the state say transportation gridlock must be relieved through a comprehensive transportation plan with adequate funding to include light rail, bike routes, and walking trails all interconnected in some way.

Legislative planning of bills and funding for transportation needs as described below have helped change the landscape for Atlanta and the region.

## ❖ House Bill 1218

Proposed legislation by governor Sonny Perdue under the Georgia General Assembly (House Bill) HB1218, entitled the Transportation Investment Act of 2010, is a bill to provide for a short title; to amend Title 32 of the O.C.G.A (Official Code of Georgia, Annotated)., relating to highways, bridges, and ferries, so as to provide for certain powers and duties of the

---

241    http://www.geogiaencyclopedia.org/nge/Article.jsp?id=h-2207

Department of Transportation; to amend Title 48 of the O.C.G.A ., relating to revenue and taxation, so as to provide legislative finding and intent; to provide for the creation of special districts; to provide for a referendum; to provide for annual reporting; to provide for Citizen Review Panels; to amend Title 50 of the O.C.G.A., relating to state government, so as to revise certain provisions relative to the Department of Transportation's allocation of funds; to provide for related matters; to provide for an effective date; to repeal conflicting laws; and for other purposes.

## ❖ Georgia Transportation Bill

The Georgia General Assembly on April 21, 2010, voted to pass HB 277. The House passed the bill 141-29, and the Senate passed the bill 43–8. The bill would divide the state into twelve regions. A "roundtable" of local elected officials in each region, working with an appointee of the governor, would draw up a list of projects for the region. The region could then submit the list to its voters for their approval in a referendum, along with a 1 percent sales tax to fund them. The bill would next go to the desk of Governor Perdue, who stated he would have to review the final language, but still supported the concept and had worked closely on the compromise.[242]

## ❖ Connect Atlanta Plan[243]

This plan is considered the City of Atlanta's First Transportation Master Planning Process.[244]

242    http://www.ajc.com/news/georgia-politics-elections/house-senate-pass-transportation-bill

243    City of Atlanta Online; http://174.37.215.145/media/nr_masterplan_090508.aspx

244    http://web.atlantaga.gov/connectatlanta/images/Open%20House%20News%20Release.pdf

A key goal of the Connect Atlanta Plan is to complement the long-range vision and goals of the city, while focusing on specific transportation improvements for the city's residents, workers, and visitors. Providing a transportation system fit for all modes and users, regardless of physical abilities and age, is of paramount importance.

Overarching Connect Atlanta Plan goals are to encourage projects that provide for balanced transportation choices, which will help reduce dependency on single-occupant automobiles and decrease traffic congestion. In the study, transportation alternatives include public transit, bicycle lanes, and safe pedestrian walkways in high-traffic-volume areas as well as road improvements.

The Connect Atlanta Plan calls for the consideration of projects that prepare for the projected growth of the city, and include strategies that encourage fiscal and environmental sustainability for the natural and human environment. Linked to these projects are the preservation of neighborhoods and the development of more livable communities, parks, and streets through an interconnected transportation network. Extensive attention is given to alternate modes of transportation, such as:

- The addition of ninety-five miles of high-frequency transit to the existing transportation network.

- The creation of a 200-mile bicycle network that connects 95 percent of city residents to schools, parks, and activity centers.

- Safe sidewalks and street crossings that encourage pedestrian flow through neighborhoods and other community areas for work, shopping, or recreation.

### ❖ Intelligent Transportation Systems

Digital road signs on the highways and cameras on the interstates that allow Highway Emergency Response Operators (HERO) units to get to an accident quickly. The program alsoadds ramp meters on interstates.

### ❖ Congestion Management

A planning program by the Atlanta Regional Commission (ARC) that addresses the most congested corridors, the most congested roadways in the region and prioritizes them for the future investments whereby the ARC can fund those projects that really have the best impact on improving transportation in the region.

The creation of a National Infrastructure Reinvestment Bank[245] was first proposed by US senators Christopher J. Dodd and Chuck Hagel in 2007.

The Wikipedia description of the plan says that the then senator Barack Obama backed the proposed legislation in February 2008 and although he did not provide specifics about how the bank should operate, he suggested that the bank would borrow sixty billion dollars of federal funding to invest in infrastructure over ten years, while leveraging "up to 500 billion" dollars of private investment. It would invest in high-speed trains to provide an alternative to air travel, energy efficiency, and clean energy, among other kinds of public infrastructure. The bank's work will be determined by what will maximize our safety, our security, and our ability to compete, and it would create nearly two million new jobs, mainly in the construction industry. The bank would complement existing federal programs to fund infrastructure, such as the Highway Trust Fund or State Revolving Funds. It is expected to invest primarily in surface transportation infrastructure, which is likely to include highways and mass transit.

---

245    http://en.wikipedia.org/wiki/National_Infrastructure_Reinvestment_Bank

The proposed bank legislation did not mention investments in water supply and sanitation as an area of activity for the new bank. The American Water Works Association (AWWA) estimates that investment in water supply and sanitation in the United States will have to be "over $250 billion above the current levels of spending in the next 20 to 30 years" to replace aging infrastructure.

According to an article by Brad Plumer in the *Washington Post*[246] dated September 19, 2011, the proposed model would take ten billion dollars in start-up money and identify transportation, water, or energy projects that lack funding. Eligible projects would need to be worth at least one hundred million dollars and provide "a clear public benefit." The bank would then work with private investors to finance the project through cheap long-term loans or loan guarantees, with the government picking up no more than half the tab—ideally, much less—or any given project. Most US infrastructure is funded through either federal outlays or state and local municipal bonds. The country lacks a central source of low-cost financing for big construction projects, an institution akin to the European Investment Bank. The White House estimates that its infrastructure bank could ultimately backstop about one hundred billion to $200 billion dollars in construction.

## ❖ The American Reinvestment and Recovery Act[247]

The American Recovery and Reinvestment Act of 2009, signed into law in March 2010 by President Obama, is a $787 billion measure that includes $130 billion of construction spending.[248] Department of Transportation secretary Ray LaHood has set up an internal Transportation Investment Generating Economic Recovery (TIGER) team for the purpose of making

---

246    http://www.washingtonpost.com/business/economy/how-obamas-plan-for-infrastructure-bank-would-work

247    http://www2.ed.gov/policy/gen/leg/recovery/implementation.html

248    Enr.com, Feb 23, 2009

sure the Department of Transportation's portion of recovery goes out to states and localities as quickly as possible in order to immediately create jobs and strengthen our economy and transportation system. President Obama proclaimed it to be the "largest investment in our infrastructure since Eisenhower built the interstate highway system."

In January 2010, under the American Recovery and Reinvestment Act, the Department of Transportation was awarded eight billion dollars for high-speed rail grants and a plan for the DOT that calls for a four-billion-dollar National Infrastructure Innovation and Finance Fund to award grants to transportation projects that have regional and national significance.

Under the American Recovery and Reinvestment Act (ARRC), the state of Georgia was awarded $110 million for road-widening and road-resurfacing and bridge-replacement projects throughout the state.

A December 2009 report[249] found that the city had eighteen bridges that need to be repaired or replaced and the work, officials say, will cost about $162 million.

In Atlanta, these projects are proposed, in the planning stage, on hold, approved, in progress, or have been completed.

- The Mitchell Street Bridge Renovation Project in Atlanta will give an $8.3 million facelift to the nearly ninety-year-old bridge. The bridge was closed to oversized vehicles and heavy construction trucks in 2008 because it had structurally deficient beams and roadway surface. Officials deemed the bridge "unsafe and in need of repair." The new bridge will have three lanes, sidewalks, bike lanes, and a parking lane. The bridge reopened in August of 2012. The project is complete.

---

249    www.ajc.com/news/atlanta

- The Atlanta Streetcar Project is a transportation project focused on the downtown district and on travel by light rail around many of Atlanta's popular tourist attractions. The Atlanta Streetcar will be built in two phases. Phase one will focus on the east-west direction of town, and Phase Two will focus on the north-south direction of town. The project will run through the downtown tourist district, connecting Centennial Olympic Park with the Dr. Martin Luther King Jr. National Historic Site, International Boulevard, Luckie Street, Auburn Avenue, and Edgewood Avenue. The 2.6-mile Atlanta Streetcar[250] ceremonial kick-off met with a lot of fanfare on Wednesday February 1, 2012, with Mayor Reed joined by US Transportation secretary Ray LaHood and a host of city dignitaries. The streetcar will be funded through a $47.6 million federal grant, the largest federal grant allocation to the City of Atlanta for transportation other than MARTA or Hartsfield-Jackson International Airport in more than a decade. Mayor Reed said at the ceremony, "Today's launch of the Atlanta Streetcar project is the first step in a project that will transform our downtown corridor." The project is ongoing.

- The Connect Atlanta Plan is a greater transportation initiative to bring better options to intown Atlanta, like connecting attractions with the BeltLine and the Atlanta Streetcar Project. Under this plan, planning will become reality. The project is ongoing.

- The Seventeenth Street Bridge is an 830-foot-long bridge that spans twenty traffic lanes across Interstate 75/85 in downtown

---

250    http://www.bizjournals.com/atlanta/news/2012/02/01/atlanta_streetcar_project_under_way_htm

Atlanta. The 137-foot-wide, steel-box girder structure is complete with twenty-two-foot-wide sidewalks and was designed to accommodate automobiles, pedestrians, buses, bicycles, and rail. Built for $38.2 million, it links the city's Midtown office/arts/entertainment/residential district with Atlantic Station, which is a mixed-use, office/retail/residential redevelopment of the former Atlantic Steel Mill site. The project is completed.

- The Fourteenth Street Bridge Improvement Project, at a cost of eighty-eight million dollars, features two new exit ramps (southbound to Tenth Street and northbound to Seventeenth Street), new, wider, and longer bridge features dedicated turning lanes, a new tree-lined median extending from Fowler Street to West Peachtree Street, raised landscape medians, and sidewalks, and pedestrian access to Georgia Tech's Distance Learning and Professional Education Unit and the Global Learning Center midtown corridor. The bridge was closed in 2007 to relieve congestion at the Fourteenth Street/Interstate 75/85 ramp. Susan Mendheim, president and CEO, of the Midtown Alliance, said, "The Midtown Alliance could not be more complimentary of GDOT and the strong collaborative community-based partnership that has defined the 14th Street bridge project and led to a signature gateway that provides an elegant solution to balancing traffic, pedestrian safety and good urban design." The project is complete.

- The Fifth Street Bridge was a collaborative effort between the Georgia DOT and Georgia Tech. The narrow bridges at North Avenue, Fifth Street, and Tenth Street only provided for ease/west vehicular flow, and did not allow for pedestrian activity in a safe and inviting manner. With the investment in Technology

Square[251] and Centergy at Georgia Tech's downtown campus, there was an acknowledgment that the Fifth Street Bridge needed an improvement if pedestrian flow between the Institute and Midtown was going to be maximized. A study panel consisting of representatives from Centergy, Technology Square, Georgia Tech, the City of Atlanta, the Midtown Alliance, the state and the Department of Transportation was formed to study options for an enhancement of the bridge. The result was a tripling of the width of the bridge and the integration of green park areas and pedestrian-oriented sidewalks; all this was provided by a combination of Georgia and federal Department of Transportation funds. The bridge was redesigned and widened to include green space perched above sixteen lanes below Interstate 75/85 downtown. The bridge more than doubled in size, but no additional vehicle lanes were added. The project is complete.

- Paving the Way Home [252] is an historic transportation funding initiative under which all 159 counties in Georgia will receive funds for high priority local transportation improvements. Each year Georgia counties prioritize their transportation improvement needs and submit a list to the Georgia Department of Transportation (GDOT) to determine how much funding they will receive under LARP and State Aid. The planning project is ongoing.

- I-85 Express Lanes.[253] In November of 2008, the US Department of Transportation Congestion Reduction Demonstration

---

251    http://www.realestate.gatech.edu/docs/TechSquareDoD.pdf

252    http://gov.georgia/00/press/detail/0,2668,78006749_90413974_90398736,00.html

253    http:// www.georgiatolls.com/programs/i-85-express-lanes

(CRD) Program awarded a $110 million grant to support a $182 million transportation improvement project. The project is designed to provide more reliable travel times, commuter choices, and regional transit enhancements that will double the Xpress service in the I-85 corridor, support Xpress facilities throughout the region, and add more Xpress coaches. Included in the program was the conversion of a sixteen mile stretch of I-85 from Old Peachtree Road to Chamblee-Tucker Road, just south of I-285, from High Occupancy Vehicle (HOV) lanes to High Occupancy Toll, or Express Lanes. The GDOT website states that since the opening of the I-85 Express Lanes in October 2011, their useage has more than tripled. The project is complete.

- Rail Line Feasibility Studies. Georgia has been awarded 750,000 of federal money to conduct feasibility studies on Atlanta to Birmingham, and Atlanta to Chicago.

- High Speed Rail Corridor. The Georgia Department of Transportation in 2010 was working with its counterparts in Tennessee to seek thirty-four million dollars in federal funds to build a high-speed rail line linking Atlanta to Chattanooga, Tennessee.[254] The Federal Railroad Administration announced that states could apply for funding through the US High Speed Intercity Passenger Rail program created by Congress in 2009. The money would be used to plan and implement high-speed rail service along approved corridors. The Georgia DOT is working with other southeastern states on a rail line between Atlanta and Macon that would continue to Jacksonville, Florida. The third project approved as part of the funding

---

254    www.bizjournals.com/atlanta/stories/2010/08/09/daily11.html

application authorized in May 2010 calls for an in-state loop linking Atlanta, Athens, Augusta, Savannah, and Macon.

Note:

The Georgia Department of Transportation (GDOT) received a $4.1 million grant to complete a service-development plan and environmental study for the 250-mile passenger rail corridor between Atlanta and Charlotte. The GDOT is contributing $1.125 million for this phase of the project. The projects are in planning or on hold.

- The Southeast High-Speed Rail Corridor (SEHSR)[255] is a passenger rail transportation project in the United States to extend high-speed passenger rail services from Washington, DC, south through Richmond and Petersburg, and spur to Newport News in Virginia through Raleigh and Charlotte in North Carolina and connect with the existing high-speed rail corridor from Washington, DC, to Boston, Massachusetts, known as the Northeast Corridor. Since 1992, the US Department of Transportation (USDOT) has extended the corridor to Atlanta and Macon, Georgia; Greenville, South Carolina; Columbia South Carolina; Jacksonville, Florida; and Birmingham, Alabama. The project are in planning or on hold.

---

255    http://en.wikipedia.org/wiki/Southeast_High_Speed_Rail_Corridor

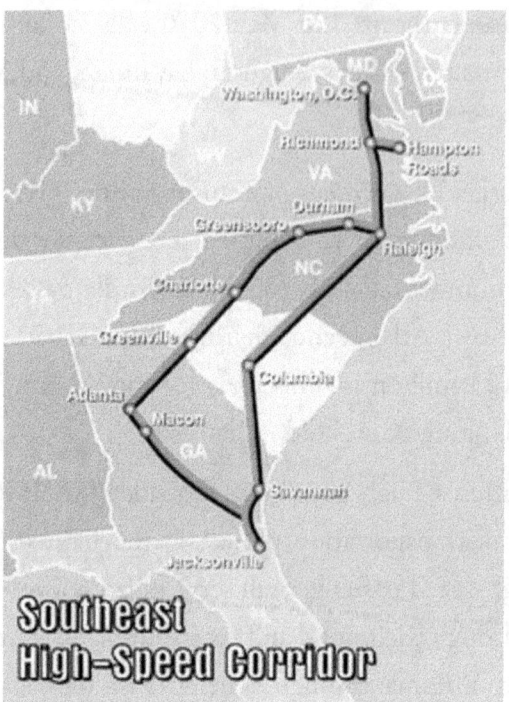

http://en.wikipedia.org/wiki/Southeast_High_Speed_Rail_Corridor

- A route from Lovejoy, Georgia. During the 2000s, Lovejoy has been proposed by the Georgia DOT to be the endpoint of metro Atlanta's first commuter rail line.[256] During the American Civil War, it was named Lovejoy Station, and was the site of the battle of Lovejoy's Station during the Atlanta Campaign in 1864. A twenty-six mile new route from Atlanta to Lovejoy, Georgia would have six future lines extended from Atlanta in all directions, to Macon, Madison, Athens, Gainsville, Canton, Bremen, and Senoia. It has eighty-seven million dollars in federal money waiting.[257] The federal dollars may disappear if state matching money is not forthcoming states the article. The project is in planning or on hold.

---

256    https://en.wikipedia.org/wiki/Lovejoy,_Georgia

257    http://www.ajc.com/opinion/atlanta-forward-a-few-63129.html

- The Northern Arc or Outer Perimeter.[258] In this article by Truman Hartshorn (a professor of geography at Georgia State University in Atlanta), he discusses issues for consideration in the public debate on the highway's fate. Metropolitan Atlanta's high rates of population and residential growth in 1990s propelled and sustained the Atlanta area to the top of the national growth charts. At the same time, the number and length of daily automobile trips skyrocketed, sprawl accelerated, and traffic congestion became the number one issue of concern to the public. Given this high level of visibility to the mobility needs of the region, it is not surprising, therefore, that the proposed Outer Perimeter became a very high-profile, if controversial and unresolved transportation question in the 1990s. Many observers, including environmentalists and citizens residing in its path, characterized the project as a worst-case example of an excessive public sector sibsidy to sprawl and pollution. Adding to the unrest, regular media reports continued to keep the issue in the public eye, polarizing opinions further. In response to the negative pressure, especially pronounced on the south side, and the uncertain status of funding sources, the Georgia Department of Transportation and the Atlanta Regional Commission began scaling back the scope of the project at the close of the decade (1990s). In place of the 200-mile circumference route, a dramatically modified Northern Arc emerged as an alternative of the Regional Transportation Plan released in the spring of 1999 by the Atlanta Regional Commission. In the summer of 1999, the Georgia Department of Transportation held a series of hearings on the proposed fifty-nine mile routeextending

---

258    http://aysps.gsu.edu/publications/researchatlanta/The%20Northern%20Arc.pdf

from I-75 in the Cartersville area eastward to I-85 and Georgia
Route 316 in the Lawrenceville area. The route would have
gone through or near the following communities: Cartersville,
Canton, Cumming, Buford, Dacula, Loganville, Conyers,
McDonough, Hampton, Newnan, Peachtree City, Villa Rica,
and Dallas.[259] The Outer Perimeter was an expressway originally planned to encircle Atlanta about twenty to twenty-five
miles farther away from the city than the existing Perimeter
Highway (I-285). The Northern Arc was to have been a toll
road under another proposal, which advocates said would have
kept local traffic away from the highway, while freeing it from
trucks. Opponents said that despite the toll, the road would
encourage additional development and congestion, creating
the continued urban sprawl that, at times, threatened to overwhelm areas much closer to Atlanta proper. The project is either on hold or has been abandoned.

---

259    http://en.wikipedia.org/wiki/Outer_Perimeter

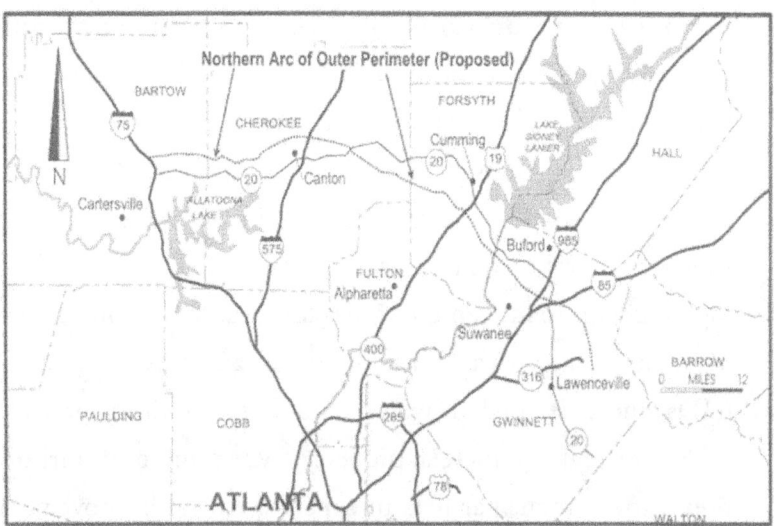

Google- The Northern Arc: The Outer Perimeter- Reincarnated?

- In 2009, MARTA Transit would receive $55.4 million in federal stimulus funds[260] from the US Department of Transportation. MARTA said it would use the money to obtain eighteen new fourty-foot buses. It would also go toward preventative maintenance, rail stations and stops, and other enhancements. In 2010, the Metropolitan Atlanta Regional Transit Authority was awarded teo million dollars in Stimulus money for operational purposes under the American Recovery and Reinvestment Act (ARRC).

Note:

MARTA will also have projects done under the Concept 3 Plan for regional transportation.

- Diverging Diamond Interchange[261] (DDI). At a cost of $4.6 million by the Georgia DOT, the redesign of the busy I-285/

---

260    http://www.ajc.com/news/marta-to-get-55.html

261    www.perimetercid.org/canyouddi/what-is-the-ddi.html

Ashford-Dunwoody Road Interchange is to improve traffic flow and safety. On June 4, 2012, metro Atlantans experienced Georgia's first newly constructed Diverging Diamond Interchange, opening to crossover traffic at I-285/Ashford-Dunwoody Road interchange near Perimeter Mall[262]. The perimeter Community Improvement Districts (PCIDs) are leading the charge to implement vital transportation enhancements. This partnership with Central (Dekalb) and Fulton Perimeter CIDs, the cities of Dunwoody and Sandy Springs, state, local and federal commercial property owners and multijurisdictional government partners to expedite critical improvements in transportation infrastructure and quality of life are aimed at relieving traffic congestion, improving choices and access and encouraging connectivity. In a DDI, the GDOT states this proven, innovative, low-cost design, cutting edge plan shifts the flow of traffic to the opposite side of the road to reduce points of traffic conflict. The plan provides immediate traffic relief and delays in evening rush hours up to 20 percent. The PCID and the GDOT say this could become the model for congested interchanges throughout the state. The project is complete.

---

262    www.perimetercid.org

*Notes :*

The Georgia Department of Transportation has put out a list of projects it anticipates starting construction on in Atlanta, if the projects are approved by the voters for upcoming years 2013–2015:

Tenth Street from Howell Mill Road to Monroe Drive- traffic improvements

Fourteenth Street from Howell Mill Road to Piedmont Road- traffic improvemnts

Auburn Avenue from Peachtree Street to Boulevard- traffic improvements

Courtland Street at CSX rail line and MARTA East line- bridge replacement

Atlanta BeltLine and Atlanta Streetcar Transit and Trail- downtown to northeast

- I-85/GA 400 Connector Ramp.[263] At an estimated funding cost of forty million dollars by GDOT (Georgia Department of

---

263    http://www.georgiatolls.com/asets/docs/a_i85ga400.pdf

Transportation), this two-year project started in March 2012. The project will include construction of two single-lane ramps in both directions of I-85 onto the Georgia 400 Interchange to address the lack of connection between Georgia 400 and I-85. The new ramps are projected to reduce traffic along local streets and improve travel times. As part of the project, a pedestrian trail will be constructed from Cheshire Bridge Road to Lenox Road, including a bridge across the North Fork Peachtree Creek. Until now, those travelling down Georgia 400 or I-85 southbound lanes have to navigate exiting Lenox Road, Buford Highway, and Sidney Marcus to go between these two major freeways opposite direction. This manuevering adds too much traffic to the three surface streets and creates bottlenecks during rush hour.[264] The project is under construction.

---

264    http://blog.georgiaroadgeek..com/2012/03/09/i85 ga400 ramp construction finally begins. aspx

- The Concept 3 Plan- Regional Transit Vision Plan

  On August 28, 2008, after over two years of effort, the Transit Planning Board (TPB) passed a resolution to adopt a regional concept vision transit plan called "Concept 3." The vision is a big picture, comprehensive multmodal plan that build upon thirty years of previous transit planning.[265]

  It begins with expaanding the existing MARTA heavy-rail system at its core and extends this sytem in the northeast, southeast, and western directions. Additionally, an inner core streetcar network is proposed.

  Concept 3 will also include several levels of bus service, freeway Bus Rapid Transit (BRT) systems, serving high-demand corridors with variable capacity that can respond to increasing demand.

  The final element of Concept 3 proposes creating a support bus network, which includes arterial rapid bus and cross-country bus services, activity-center circulators, and expanded local bus service.

  Name Change.[266] Concept 3 to Action Plan 3.0 in 2012.

  In July 2012, working to beat a midnight deadline, elected officials voted to approve a draft list of transportation projects for the ten county region. The list blends $6.14 billion of public transportation and road projects. First, it allocates significant funding for three major, new, fixed-guideway rail lines in the region that include:

---

265    http://www.ncppp.org/publications/TransitDallas_0810/B.SCOTT%20-%20TPB%20 Concept%203%20Fact%20Sheet.pdf

266    http://livecomm.wordpress/2011/08/23/from concept 3 to action plan 3 0 a new transit vision for metro Atlanta/

- More than $600 million for Atlanta's BeltLine streetcar network.

- More than $856 million for a rail line from Midtown Atlanta to Cobb County's Cumberland district.

- $700 million to connect MARTA's Lindberg station to the huge Emory University/Centers for Disease Control job center in northwest Dekalb County.

In addition, the list provides $600 million for MARTA capital "state of good repair" projects. This money not only refurbishes the region's core transit rail system but also protects access to future federal funds for other transit projects by showing good stewartship of existing investments, a key benchmark for new federal funding. The project has been approved.

The final element of Concept 3 proposes creating a support bus network, which includes arterial rapid bus and cross-country bus services, activity-center circulators, and expanded local bus service.

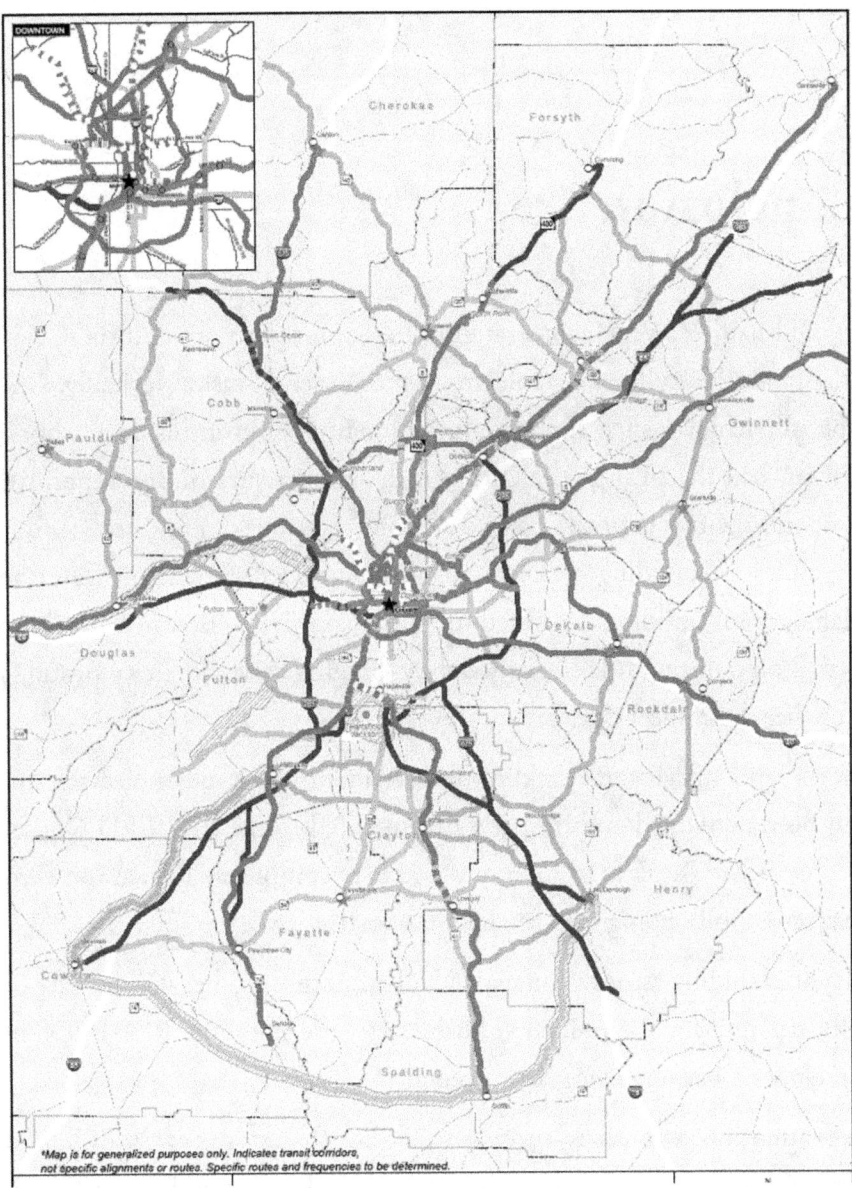

http://www.ncppp.org/publications/TransitDallas_0810/B.SCOTT%20-%20TPB%20
Concept%203%20Fact%20Sheet.pdf

# Conclusion

I would like to give Mayor Franklin, her administration, and the City of Atlanta all the kudos for achieving a monumental task. Normally, when a person achieves such success, a monument is built in his or her honor. Having said that, I urge the City of Atlanta's new mayor, his administration, and the city council's Call to Action to get busy to find source funding and start building a new state-of-the-art Department of Watershed Management building dedicated to Shirley Franklin, "Best in Class," with a large oil painting of her in the lobby. She has served the city proud, so show her the respect due.

By the end of Mayor Franklin's first term in office, all the sewer spills had been contained and properly documented to the EPA/EPD Consent Decree order. The City of Atlanta was able to completely reduce and eliminate sewer spills going into the Chattahoochee River.

Mayor Franklin, "the sewer mayor," had accomplished her vision to fix our infrastructure for the residents, businessess, the state, the EPA/EPD, and the Upper Chattahoochee Riverkeeper.

The attention was then turned to SSES (Sanitary Sewer Evaluation Surveying) by sewer group, worst (SG-1) to least critical (SG-6), along with funding to finance the work, which was a large part of assessing and repairing Atlanta's aging sewer system.

As stated in the "What's the Cost" chapter, you can see that a large emphasis was placed on SSES maintenance of Atlanta's aging sewer system.

Public information managers and officers were assigned areas throughout the city to address residents' concerns and complaints for public-relations purposes.

By the time I departed from the SSES program in November 2010, Sewer Group 5 (SG5) contracts were being issued to address the remainder of the Consent Decree order for 2014.

Mayor Franklin was leaving office in 2010 quietly and not with a lot of fanfare, and her successor, the newly elected mayor, Kasim Reed, would take the helm in January 2011.

Before Mayor Franklin departed, the state was awarded $85.5 million in projects for eight Georgia communities[267] from GEFA as part of Governor Perdue's vision for sustainable economic development for water and sewer infrastructure projects. From that, the City of Atlanta was approved for a Clean Water State Revolving Fund (CWSRF) loan of seven million dollars to finance upgrades to the R.M. Clayton Wastewater Treatment Plant and another Clean Water Revolving Fund (CWSRF) for forty million dollars and a Georgia Fund Loan of three million dollars to finance new sewer lines and sewer rehabilitation.

Mayor Franklin followed the First Amended Consent Decree (FACD) to the letter, line by line, and achieved completion by 2012, two years ahead of schedule as promised.

The City of Atlanta completely assessed its sewers with the SSES program, filled approximately 3,700 potholes; increased capacity for sewer flow by separating combined sewer basins; eliminated the worst-performing CSOs; reduced the number of permitted wet-weather overflows to a minimum of four per year; updated poorly performing wastewater-treatment facilities that went neglected; constructed a deep-rock storage tunnel; and

---

267    http://www.georgia.gov/00/press/detail/0,2668,78006749_161911047_162554339,00.
html

built state-of-the-art wastewater-treatment facilities that screen, disinfect, chlorinate and dechlorinate waste flow. The city improved the drinking-water quality for Atlantans; had a state-of-the-art water/wastewater testing laboratory built; purchased an abandoned quarry to built a deep water-well reservoir and park; improved the stream buffers throughout Atlanta creeks; conformed to NPDES standards along the creeks and tributaries; improved streets with a streetscape program and quality-of-life program; purchased land to develop greenspace for residents throughout the city; and put a bookmark in place for future projects.

Mayor Franklin led the charge to bring life back to Piedmont Park in the downtown area, and every other park throughout the city. She developed a department of watershed management with a commissioner to oversee all programs, protected the present and future water sources for the region, and created a structure that protects public health and the environment. She kept the EPA/EPD abreast of the city's progress, and from day one to the last day in office, she constantly and persistantly requested funding from anyone and everyone that could contribute.

Along the way, she won numerous awards for herself, and a state-of-the-art laboratory facility at Utoy Creek was built.

She added a new 911 emergency response/customer call center, attacked crime that was dominating the nightly news and taking over the city, canceled a multimillion-dollar contract for drinking-water services, enhanced the economic vitality and prosperity of the city with public-private partnerships, and restored confidence back to city government.

As it related to the power of partnerships,[268] Mayor Franklin stated in the article there are three things in a partnership:

---

268    http://economyleague.org/node/213

*Strategic focus*- don't take on everything at once. She focused on three goals: balance the budget, fix the sewers, and pass ethics-reform legislation. For her first few years, she talked about nothing else.

*Private sector partnerships*- She credits the private sector as having played a major role in all her successess. The city has a history of public-private partnerships going back a hundred years. There is hardly anything that we've done in Atlanta successfully that has not been a joint effort with the private sector. The process is to involve the private sector early. Typically, she would create working groups and ask them to consider possible fixes to a given problem. Then when it's time to raise money and implement a solution, she would have strong support in place. "Every time it's successful."

*High standards*- She stated, "Don't lower them." Set your target based on what you really need to get it done. And then rally those people that have a big enough view of the world to come to your aid so you can add to it incrementally.

She led by example, and people followed her to the end of her administration.

Commissioner Rob Hunter would stay on through Mayor Franklin's administration until he resigned in 2010 and stayed on as a technical adviser till December 2011. Under Rob Hunter, the Department of Watershed Management completed the Consent decree requirements on time and under budget, issued over two billion dollars in bonds and awarded fifty million dollars in state revolving-fund loans. Rob Hunter worked closely with the Mayor and the city council to develop and implement the 2004 and 2008 water and sewer rate increase packages and sales tax (MOST) referendums, which will enable the Department of Watershed Management to continue funding an operational budget in excess of $500 million and capital improvements and rehabilitation for the drinking-water and wastewater systems.

Montgomery Watson/Khafra team would stay on to serve as program manager (PMT) for the capital improvement program and provide management consulting regulatory compliance, planning, operations, engineering, and construction administration support services from 2001 to present.

What has Atlanta learned?[269]

The plans that are required under the Clean Water Atlanta program guide the city through a logical progression of data collection and evaluation that, in turn, allows the Department of Watershed Management to make informed rehabilitation and replacement decisions. The success that has been experienced to date in this program assures that the processes will result in Atlanta having a first-class water and wastewater collection, transmission and treatment infrastructure by 2014. Atlanta's Clean Water program is one of the highest-profile wet-weather control programs of its kind in the nation.

When City officials are deciding on an infrastructure project, a Sanitary Sewer Evaluation Survey study will give the most valuable information on the process, techniques, and tools used, and most importantly a cost of some of the activities involved. Officials should be able to prioritize maintenance money through capital improvement budgeting to get the best value for dollars spent or return on investment (ROI).

Assessing the need for infrastructure rehabilitation and repair across the country is vital to developing an Integrated Sustainable Project Lifecycle approach. Cities that need this repair should already have projects "shovel-ready." Cites that are not doing infrastructure work should be discussing with cities like Atlanta to see how they are becoming EPA-compliant and taking a leadership role in defining sustainable solutions because repairing

---

269    City of Atlanta, Dept. of Watershed Management (Asset Management Case Study), What has Atlanta learned?

and maintaining aging sewers is not a short-term fix and planning needs to be done for five to ten years.

By using the repair methods described, maintaining and inspecting sewer pipes and structures on a routine maintenance schedule, pipes' life expectancy can increase by twenty or more years. As the years go by, new equipment and advanced techniques are replacing old ones.

The Department of Watershed Management for the City of Atlanta has sucessfully separated a combined storm and sewer system and has established and implemented a sanitary sewer evaluation survey rehabilitation program that maintains the system. The department has contracted with qualified contractors to use state-of-the art repair techniques to repair an aging system that will become a model system for all cities to emulate.

Along the way, throughout Mayor Franklin's administration a lot of plans, programs, and initiatives were created out of the need to aid infrastructure projects either directly or indirectly with the intent of complying with Consent Decree-related items. This helped shape the blueprint for policy, strategy, and implementation towards the mayor's vision for "Best in Class."

A model for infrastructure rebuilding[270]

In infrastructure terms, Clean Water Atlanta has become a twenty-first century model for water and sewer system rebuilding. The program has resulted in cleaner rivers and streams, has allowed development to proceed, and has been accomplished on time and budget despite oppressively tight deadlines.

On another note, the Associated Builders and Contractors' Construction Backlog Indicator shows infrastructure is the sector with the healthiest amount of work, with the House of Representatives passing a

---

270    City of Atlanta Online http://www.atlantaga.gov/media/nr_cleanwater_080509.aspx

$154 billion "Jobs for Main Street," with forty-seven billion dollars going to infrastructure.

From a global perspective, we see that America lags far behind other countries in terms of scope and dollars invested to upgrade its infrastructure (see India).

Congress is doing its best to try to keep up with the demand cities and states are putting on them for funding, but its not enough. As individual cities across the country, we need to do more to put money into the hands of those officials that can put people to work, all the while stimulating the economy.

As far as transportation is concerned, Atlanta and the surrounding ten counties have positioned themselves to be *THE* premier leader in transportation for the state with the help of the Georgia General Assembly, state legislators, county commissioners, mayors, ARC, DCA, GRTA, GDOT, MARTA, TPB partners, and various governmental and business groups with the funding from federal transportation.

Lastly, if your city is cited by the EPA as noncompliant regarding its sewers/water and has to enter into a consent decree, do not look at it as a "death wish," but consider it an opportunity to move your City forward into the future.

Follow the example of Atlanta, Mayor Franklin and her administration.

Turn those lemons into lemonade. Look what happened here!

My hope is that the information provided to you here is informative and that it will give you an understanding into what goes on with infrastructure work in rebuilding one of America's great cities.

Now you will be able to tell officials, "Fix Our Infrastructure."

# Recommended websites:

- ➤ Underground Construction- www.ucononline.com
- ➤ Engineering News- Record (ENR) magazine- www.enr.construction.com
- ➤ American Society of Civil Engineers (ASCE)- www.asce.org
- ➤ Governing magazine- www.governing.com
- ➤ Sustainable Business- www.sustainablebusiness.com
- ➤ Journal AWWA (American Water Works Association)- www.awwa.org
- ➤ Associated General Contractor (AGC)- www.agc.com
- ➤ Infrastructure Investor- www.perimedia.com/infrastructureinvestor.com
- ➤ Energy and Infrastructure- www.energyandinfrastructure.com
- ➤ InfraStructures- www.infrastructure.com
- ➤ World Highways- www.ropl.com/magazines/world-highways
- ➤ American Infrastructure- www.americainfra.com
- ➤ Roads And Bridges- www.roadsbridges.com
- ➤ Better Roads- www.betterroads.com
- ➤ Accelerated Bridge Construction- www.structuremag.org
- ➤ MassTransit- www.masstransitmag.com
- ➤ American Public Transportation Association- www.apta.com
- ➤ The Transport Politic- www.thetransportpolitic.com